Birkhäuser

Frontiers in Mathematics

This series is designed to be a repository for up-to-date research results which have been prepared for a wider audience. Graduates and postgraduates as well as scientists will benefit from the latest developments at the research frontiers in mathematics and at the "frontiers" between mathematics and other fields like computer science, physics, biology, economics, finance, etc. All volumes are online available at SpringerLink.

Sorin G. Gal • Constantin P. Niculescu

Choquet Integral and Monotone Sublinear Operators

Birkhäuser

Sorin G. Gal
Oradea, Romania

Constantin P. Niculescu
Department of Mathematics
University of Craiova
Craiova, Romania

ISSN 1660-8046 ISSN 1660-8054 (electronic)
Frontiers in Mathematics
ISBN 978-3-031-71700-0 ISBN 978-3-031-71701-7 (eBook)
https://doi.org/10.1007/978-3-031-71701-7

This book is published under the imprint Birkhäuser, www.birkhauser-science.com by the registered company Springer Nature Switzerland AG
The registered company address is: Gewerbestrasse 11, 6330 Cham, Switzerland

Preface

The notion of measure (a natural generalization of the concepts like length, area, and volume) plays an important role in mathematics and has deep implications in science, engineering, economics, etc. In the early twentieth century, the development of the theory of Lebesgue measure and Lebesgue integral laid the foundations for modern probability theory, stochastic processes, and the study of partial differential equations, all based on the methods of functional analysis.

The last 50 years have shown a serious change in our understanding of the notion of measure. Indeed, the progress of research in a series of fields (like potential theory, cooperative games, decision theory under risk, decision theory under uncertainty, and discrete convex analysis) revealed that the property of additivity (so basic in the theory of Lebesgue measure and of Lebesgue integral) is sometimes too restrictive. This led to the consideration of some classes of nonnegative set functions, called capacities (or fuzzy measures), playing the role of "nonadditive measures."

A capacity is subject to very few restrictions: the capacity of the empty set is 0, and if a set A is included in a set B,, then the capacity of A does not exceed the capacity of B. The theories of integration associated with capacities proved to be unexpectedly rich. In particular, so is the case of the Choquet integral, which plays a central role in this book.

Just to offer an idea of what it is about, let us consider the case of a street vendor offering three distinct products a_1, a_2, and a_3 at the prices per unit, respectively, $\mu(\{a_1\})$, $\mu(\{a_2\})$, and $\mu(\{a_3\})$. Denote $\Omega = \{a_1, a_2, a_3\}$. The options for the customer are

$$\emptyset \text{ (no choice)}, \{a_1\}, \{a_2\}, \{a_3\},$$

$$\{a_1, a_2\}, \{a_1, a_3\}, \{a_2, a_3\} \text{ and } \{a_1, a_2, a_3\}.$$

Mathematically, this list of possible orders is nothing but the power set $\mathcal{P}(\Omega)$ of Ω and it comes with a price offer $\mu : \mathcal{P}(\Omega) \to [0, \infty)$, that assigns a price to each choice. The simplest formula is to cash the products independently, i.e.,

$$\mu(A) = \sum_{a_k \in A} \mu(\{a_k\}).$$

In this case, $\mu(A \cup B) = \mu(A) + \mu(B)$ if A and B have no common elements (so μ is an additive measure). Instead, if a promotion like $\mu(\{a_1, a_2, a_3\}) < \mu(\{a_1\}) + \mu(\{a_2\}) + \mu(\{a_3\})$ is offered, then the additivity of μ is broken, so that μ is merely a capacity. If someone buys $f(a_1)$ pieces of a_1, $f(a_2)$ pieces of a_2, and finally $f(a_3)$ pieces of a_3 and $f(a_1) \leq f(a_2) \leq f(a_3)$, then the entire purchase costs

$$f(a_1)\mu(\{a_1, a_2, a_3\}) + (f(a_2) - f(a_1))\mu(\{a_2, a_3\}) + (f(a_3) - f(a_2))\mu(\{a_3\}).$$

Understanding the functions $f : \Omega \to \mathbb{N}$ as possible purchases and denoting their costs as (C) $\int_\Omega f \, d\mu$, we get a suggestive encounter with the Choquet integral.

The basic elements of Choquet integral were introduced in 1925 by G. Vitali [178], but his work went more or less unnoticed at that time. Marinacci [123] has published a short note on this subject.

The true birth of the theory of capacities and the integrals associated with capacities is marked by Choquet's seminal paper [41] from 1953. His original goal was to use capacities and Choquet integrals in statistical mechanics and potential theory. A historical view of that moment was offered by Choquet himself in [42]. Soon after, capacities and their associated integrals started to attract economists' attention. The connection with decision theory was explicitly recognized with the work of Dempster [50], later developed by Shafer [166]. The capacities and Choquet integrals were rediscovered in 1986, when David Schmeidler [164, 165] put forward an axiomatic model of choice with nonadditive beliefs. Continuing with the contributions of Gilboa and Schmeidler [86] and Gilboa [85], the theories of decision-making under risk and uncertainty have undergone a substantial development in finance, economics, insurance, and machine learning.

Nowadays, the whole range of applications of the Choquet integral has become so vast that we cannot get into further details due to our limited space.

However, we will mention some of the many fine books on these topics: Aumann and Shapley [17], Adams and Hedberg [3], Denneberg [51], Gilboa [85], Grabisch [90], Landkof [112], Ponce [150], Torra et al. [176], and Wang and Klir [183].

The purpose of the present book is not to focus on the traditional applications of Choquet integral. Instead, it is devoted to the study of a class of operators that combines the properties of subadditivity, monotone increasing, and a restricted form of translation invariance, all very known in the case of Choquet type operators (i.e., of operators defined by Choquet integrals associated with vector capacities). The large list of such operators includes, for example, all linear and positive operators $T : C([0, 1]) \to C([0, 1])$ such that $T(1) = 1$ (and the same is true if $C([0, 1])$ is replaced by one of the Lebesgue spaces $L^p([0, 1])$ with $p \in [1, \infty]$). It also includes two curious examples of nonlinear functionals from Real Analysis, such as $U : \ell^\infty \to \mathbb{R}$, defined by $U((x_n)_n) = \limsup_{n\to\infty} x_n$, and $V : \ell^\infty \to \mathbb{R}$, defined by $V((x_n)_n) = \sup_{n\in\mathbb{N}} x_n$. See Sect. 4.1 for details and more examples. The interest in these operators is motivated in Chap. 5, where they provide the framework for numerous nonlinear extensions of the classical Korovkin theory of approximation.

The intended reader we have in mind should have a graduate course on Real analysis and some knowledge of Functional Analysis, but no previous knowledge about capacities and the Choquet integral is required.

Chapter 1 is devoted to a quick review of some very basic definitions and results from measure theory, ordered Banach space theory, and convex analysis, all necessary for understanding the content of this book. For more, we recommend the first volume of *Measure Theory* of Bogachev [25], the book on *Real Analysis* of Carothers [32], and the *Functional Analysis* books of Bhatia [24] and Willem [187]. The main theories of integrability of functions of a real variable are covered by the textbook of Choudary and Niculescu [43], while the necessary background of Convex Analysis can be found in the book [137] by Niculescu and Persson.

Chapter 2 is devoted to a thorough presentation of the notions of capacity and Choquet integral. Sections 2.1 and 2.2 deal with capacities, understood as set functions defined on an algebra of sets and taking values in $[0, \infty]$. They include numerous examples of capacities enjoying various special properties like continuity (from above or from below) and strong subadditivity/strong superadditivity. The algebras of sets are instrumental for defining the measurable functions in Sect. 2.3. The construction of the Choquet integral in Sect. 2.4 follows the historical route. First, for the nonnegative measurable functions (as Choquet himself did), and then by extending it to the case of measurable functions with variable sign, via the property of invariance to translations. Note that the Choquet integral coincides with the Lebesgue integral in the case of σ-additive measures defined on σ-algebras of sets. The crucial property of comonotonic additivity of the Choquet integral (as well as some immediate applications) is the subject of Sect. 2.5. Another important result concerning the Choquet integral is the equivalence between its subadditivity and the strong subadditivity of the underlying capacity. Choquet capacities and Choquet integrals are known to be difficult to compute. Some partial results, based on the work of Sugeno [173, 174] and Ridaoui and Grabisch [157], were incorporated in Sect. 2.7. Chapter 2 continues with two very important results: Schmeidler's representation theorem and the analog of the Riesz-Kakutani representation theorem for capacities. The Comments section includes (among many other remarks) a discussion about games and the way the games with bounded variation generate integrals à la Choquet.

Chapter 3 is devoted to the capacitary analogs of the Lebesgue L^p-spaces for $p \in [1, \infty]$. Here we combine ideas due to Adams [2], Cerdà et al. [34], and Ponce and Spector [151]. A special attention is paid to the extension of the classical inequalities (like the modulus inequality, Chebyshev inequality, Hölder inequality and Minkowski inequality) to the framework of capacities. Another important auxiliary result is Fatou's lemma, whose connection with decreasing rearrangements is exposed following the approach of Cerdà et al. [34]. The resulting capacitary L^p-spaces are Banach lattices in the case of lower continuous, strongly subadditive and finite capacities defined on measurable spaces. See Sect. 3.5. The case of Luzin's theorem and its connection with the approximation of functions in an L^p-space (with $p \in [1, \infty)$) by continuous functions with compact support makes the content of Sect. 3.6.

Chapter 4 is devoted to the presentation of the main properties of the class of monotone increasing weakly nonlinear operators (i.e., sublinear and translatable in the sense of Definition 4.1.1). A big source of such operators is provided by the Choquet-Bochner integrals associated to vector capacities. Formally, these integrals result by replacing the Lebesgue integral by the Bochner integral in the definition of Choquet integral. Their theory is detailed in Sect. 4.2, where it is noted that many of the nice results known in the context of Choquet integral remain available in the vector case too. So are the translation invariance, the additive comonotonicity, and the characterization of sublinearity of the integral by the strong subadditivity of the involved capacity. Extension of the representation theorem are presented in Sects. 4.3 and 4.4. Section 4.5 opens the gate for application to integral equations, describing the generalization of the Volterra operator to the context of Choquet integrals. A new generalization of the Hölder inequality in the context of monotone increasing and sublinear operators is presented in Sect. 4.6. The Comments section at the end of Chap. 4 includes a brief overview of the strongly subadditive operators and also of the submodular operators (as natural companions of certain classes of set functions).

Chapter 5 presents a series of generalizations of the Korovkin theorem in the framework of monotone increasing, sublinear and translatable operators. This theorem, which represents a clever generalization of Bernstein's famous theorem of approximation, asserts that a sequence of linear and positive operators $T_n : C([0, 1]) \to C([0, 1])$ converges to the identity of $C([0, 1])$ if this happens for a family of only three "test functions," $1,\ x$, and x^2. In Sect. 5.1, this result is generalized to the context of monotone increasing and weakly nonlinear operators defined on a vector lattice of functions (defined on a locally compact subset Ω of the Euclidean space $\mathbb{R}^N$) that contains the $2N + 2$ the test functions $1,\ \pm \mathrm{pr}_1, \ldots, \pm \mathrm{pr}_N$ and $\sum_{k=1}^{N} \mathrm{pr}_k^2$; here pr_k denotes the projection on the component of order k. See Theorem 5.1.1 and its illustrations in the case of Bernstein-Kantorovich-Choquet operators, Szász-Mirakjan-Kantorovich-Choquet operators, and Baskakov-Kantorovich-Choquet operators. The case of separating functions as a substitute for the test functions is discussed in Sects. 5.2 and 5.3. The next three sections are devoted to the Korovkin type theorems concerning the almost everywhere convergence, the convergence in measure, and the convergence in L^p-norm. The convergence to other operators than the identity is discussed in Sect. 5.7 (accompanied by the quantitative estimates proved in Sect. 5.8). Finally, in Sect. 5.9, the classical probabilistic scheme of Feller is extended to the case of the Choquet integral, generating new examples of monotone increasing and sublinear operators with good approximation properties.

The book is mainly addressed to graduate students and researchers in domains connected with mathematics and statistics, finance, insurance, and machine learning, interested in exploring problems in these fields, marked by limited knowledge and/or uncertainty.

Supplementary material, comments, and an errata (if necessary) will be available on my web page http://math.ucv.ro/niculescu/.

Acknowledgments The manuscript of this book was finished after Sorin G. Gal passed away, leaving a preliminary draft that was meant to be further discussed, refined, and improved by the two authors together. The text underwent a thorough revision, and most parts have been completely rewritten. While this edition is my full responsibility, I wish to acknowledge the enormous debt I owe to the late Sorin G. Gal for his guidance and advice throughout our collaboration.

I am grateful to Professors Ştefan Cobzaş, Sorin Daniel Micu, Flavia-Corina Mitroi-Symeonidis, and Radu Păltănea, who read and commented on various versions and parts of the manuscript. Let me also mention the generous help offered by Lucian Coroianu, Dan-Ştefan Marinescu, Radu Miculescu, Nicuşor Minculete, and Lucian Niculescu, who contributed to the final revision of the manuscript.

I am also deeply indebted to Elizabeth Loew, Executive Editor at Springer, for her great cooperation and patience, which made it possible to successfully complete this project.

Craiova, Romania Constantin P. Niculescu

Contents

List of Symbols

$\mathbb{N}, \mathbb{Z}, \mathbb{Q}, \mathbb{R}, \mathbb{C}$	the classical numerical sets (naturals, integers etc.)
$\mathbb{N}^\star$	the set of strictly positive integers
$\mathbb{R}_+$	the set of positive real numbers
$\mathbb{R}_+^\star$	the set of strictly positive real numbers
$\overline{\mathbb{R}}$	the set of extended real numbers
$\bigvee$ and $\bigwedge$	alternative notation for sup and inf
$\complement A$	complement of A
$\subsetneq$ and $\subset$	strict and nonstrict inclusion
$\lvert A\rvert$	cardinality of A
$\emptyset$	empty set
χ_A	characteristic function of A
I_A	identity of the set A
$1\lvert_A$	function identically 1 on A
$f\lvert_A$	restriction of f to A
$\liminf_{x\to a}$	limit inferior
$\limsup_{x\to a}$	limit superior
$\omega(f,\delta)$	modulus of continuity
∂A	boundary of A
$\overline{A}$	closure of A
$\operatorname{int} A$	interior of A
$B_r(a)$	open ball with center a and radius r
$\overline{B}_r(a)$	closed ball with center a and radius r
$\operatorname{diam}(A)$	diameter of A
$d_U(x) = d(x,U)$	distance from x to U
$\mathbb{R}^N$	Euclidean N-space

$\mathbb{R}_+^N = \{(x_1,\dots,x_N) \in \mathbb{R}^N : x_1,\dots,x_N \geq 0\}$, the positive orthant
$\mathbb{R}_{++}^N = \{(x_1,\dots,x_N) \in \mathbb{R}^N : x_1,\dots,x_N > 0\}$
$\mathbb{R}_{\geq}^N = \{(x_1,\dots,x_N) \in \mathbb{R}^N : x_1 \geq \dots \geq x_N\}$

$\langle x, y\rangle, x \cdot y$	inner product
$\dim E$	dimension of the space E
C^*	dual cone
E^*	dual space
$[x, y]$	line segment
$\mathrm{conv}(A)$	convex hull of A
$\overline{\mathrm{conv}}(A)$	closed convex hull of A
$\mathrm{span}(A)$	linear span of A
$\overline{\mathrm{span}}(A)$	closed linear span of A
$\lambda A + \mu B = \{\lambda x + \mu y : x \in A,\ y \in B\}$	Minkowski operations
$\mathrm{M}_N(\mathbb{R}),\ \mathrm{M}_N(\mathbb{C})$	spaces of $N \times N$-dimensional matrices
$\mathrm{Sym}(N, \mathbb{R})$	the space of all symmetric matrices of $\mathrm{M}_N(\mathbb{R})$
$\mathrm{Sym}^+(N, \mathbb{R})$	the subset of all positive matrices of $\mathrm{Sym}(N, \mathbb{R})$
$\mathrm{Sym}^{++}(N, \mathbb{R})$	the subset of all strictly positive matrices of $\mathrm{Sym}(N, \mathbb{R})$
I	identity matrix
A^*	adjoint matrix
$P_K(x)$	orthogonal projection
$\det A$	determinant of A
$\mathrm{trace}\, A$	trace of A
$f'_+(\mathbf{a}; \mathbf{v})$ and $f'_-(\mathbf{a}; \mathbf{v})$	directional derivatives
$\partial_{\mathbf{a}} f$	subdifferential of f at $\mathbf{a}$
$f'(\mathbf{a}; \cdot)$	Gâteaux derivative at $\mathbf{a}$
∇	gradient
Hess, ∇^2	Hessian matrix
$\mathrm{d}f(\mathbf{a})$	Fréchet differential at $\mathbf{a}$
$\mathcal{F}(\Omega)$	space of real-valued functions defined on Ω
$\mathcal{F}_b(\Omega)$	space of bounded real-valued functions defined on Ω
$C(\Omega)$	space of real-valued continuous functions
$C_b(\Omega)$	space of real-valued continuous and bounded functions

$C_c(\Omega) = \{f : f \in C(\Omega),\ f \text{ having compact support}\}$
$C_0(\Omega) = \{f : f \in C(\Omega), \text{ continuous and vanishes at infinity}\}$
$\mathcal{U}C(\Omega) = \{f : f : \Omega \to \mathbb{R} \text{ uniformly continuous}\}$
$\mathcal{U}_bC(\Omega) = \{f : f : \Omega \to \mathbb{R} \text{ uniformly continuous and bounded}\}$
$C^m(\Omega) = \{f : D^\alpha f \in C(\Omega) \text{ for all } |\alpha| \le m\}$
$C^m(\overline{\Omega}) = \{f : D^\alpha f \text{ uniformly continuous on } \Omega \text{ for all } |\alpha| \le m\}$

$C_c^\infty(\Omega)$	space of functions of class C^∞ with compact support
$\mathcal{M}(\Omega, \mathcal{A})$	set of measurable functions
$\mathcal{M}_b(\Omega, \mathcal{A})$	set of bounded measurable functions
$\mathcal{M}_b^+(\Omega, \mathcal{A})$	set of positive and bounded measurable functions
$\mathcal{S}(\Omega, \mathcal{A})$	space of $\mathcal{A}$-measurable simple functions defined on Ω
ℓ^p	space of p-th-power absolutely summing real sequences
$L^p(\Omega)$	space of p-th-power Lebesgue integrable functions on Ω
$L^p(\mu)$	space of p-th-power μ-integrable functions
a.e	almost everywhere
q.e.	quasi-everywhere
q.u.	quasi-uniformly
$f * g$	convolution
$\\|f\\|_{L^p}$	L^p-norm
$bv(\mathcal{A})$	set of games with finite variation
$\\|\nu\\|_{bv}$	variation norm
$\mathrm{Lip}(f)$	Lipschitz constant
$\mathcal{B}(\Omega)$	Borel σ-algebra associated to Ω
$\mathrm{Prob}(\Omega)$	set of Borel probability measures on Ω
δ_a	Dirac measure concentrated at a
m	Lebesgue measure
$\mathrm{Vol}_N(K)$	N-dimensional volume of K
$\mathrm{bar}(\mu)$	barycenter of μ
$\mathrm{E}(f)$	expected value (mean value) of f
$\mathrm{Var}(f)$	variance of f

■ end of the proof

Preliminaries 1

The topics discussed in this book require some previous knowledge of Real Analysis, Measure Theory, Functional Analysis, and Convex Analysis. For the convenience of our readers, we will recall here a number of very basic facts. For more, we recommend the first volume of *Measure Theory* of Bogachev [25], the book on *Real Analysis* of Carothers [32], the *Functional Analysis* books of Bhatia [24] and Willem [187], and the *Introduction to Probability and Statistics* by Dekking et al. [48]. The main theories of integrability of functions of a real variable are discussed in the textbook of Choudary and Niculescu [43], while the necessary background of Convex Analysis is covered by the book [137] of Niculescu and Persson.

1.1 Measure Theory

One of the main concepts of measure theory is that of measure space. We start its description by mentioning several classes of sets.

Definition 1.1.1 Given a nonempty set Ω, a *lattice* of Ω is any nonempty family $\mathcal{F}$ of subsets of Ω that is closed under finite intersections and finite unions. If, in addition, $\mathcal{F}$ is closed under difference, then $\mathcal{F}$ is called a *ring*.

An *algebra* is any ring $\mathcal{A}$ of subsets of Ω that contains the "total set" Ω.

An algebra Σ closed under countable unions and countable intersections is called a *σ-algebra*.

Any pair (Ω, Σ) consisting of a set Ω and a σ-algebra Σ of subsets of Ω is called a *measurable space*.

S. G. Gal, C. P. Niculescu, *Choquet Integral and Monotone Sublinear Operators*,
Frontiers in Mathematics, https://doi.org/10.1007/978-3-031-71701-7_1

Necessarily, any ring contains the empty set (so every algebra of Ω contains at least the sets $\emptyset$ and Ω).

The *power set* $\mathcal{P}(\Omega)$ of a set Ω (i.e. the set of all subsets of Ω) is an example of σ-algebra. So is $\{\emptyset, \Omega\}$. When Ω is a Hausdorff topological space, one can speak of the *Borel σ-algebra* of Ω, the σ-algebra $\mathcal{B}(\Omega)$ generated by the open sets (i.e., the intersection of all σ-algebras $\mathcal{A}$ included in $\mathcal{P}(\Omega)$ that contain the open subsets of Ω). Equivalently, $\mathcal{B}(\Omega)$ is the smallest σ-algebra of subsets of Ω containing the open sets. The elements of a Borel σ-algebra are referred to as *Borel sets.*

Clearly, $\mathcal{B}(\Omega)$ also contains all closed subsets of Ω, all G_δ subsets (i.e., the countable intersections of open sets) and all F_σ subsets (i.e. the countable unions of closed sets). The Borel σ-algebra of $\mathbb{R}$, $\mathcal{B}(\mathbb{R})$, is the smallest σ-algebra on $\mathbb{R}$ that contains all the intervals.

Remark 1.1.2 The Borel σ-algebra $\mathcal{B}(\mathbb{R})$ is generated by each of the following classes of sets:

(i) the family of all intervals;
(ii) the family of all intervals with rational endpoints;
(iii) the family of all rays of the form $(-\infty, c)$, where $c \in \mathbb{R}$;
(iv) the family of all rays of the form $(-\infty, c]$, where $c \in \mathbb{R}$;
(v) the family of rays of the form $(c, +\infty)$, where $c \in \mathbb{R}$;
(vi) the family of all rays of the form $[c, +\infty)$, where $c \in \mathbb{R}$.

Finally, the same is true if in place of all real numbers c, one takes points of any everywhere dense subset of $\mathbb{R}$.

Two important lattices that can be attached to a compact Hausdorff space Ω are: $\Sigma_{up}^{+}(\Omega)$, the lattice of all *upper contour closed sets*

$$S_{f,t} = \{x \in \Omega : f(x) \geq t\},$$

and the lattice $\Sigma_{up}^{-}(\Omega)$ of all *upper contour open sets* $S'_{f,t} = \{x \in \Omega : f(x) > t\}$, associated with pairs $f \in C(\Omega)$ and $t \in \mathbb{R}$.

When Ω is a compact metrizable space (with the metric d), $\Sigma_{up}^{+}(\Omega)$ coincides with the lattice of all closed subsets of Ω (and $\Sigma_{up}^{-}(\Omega)$ coincides with the lattice of all open subsets of Ω). Indeed, if $C \subset \Omega$ is a closed subset, then $C = \{x \in \Omega : -d(x, C) \geq 0\}$.

A function whose domain is a lattice $\mathcal{L}$ of subsets of some nonempty set Ω is usually called a *set function.* Of special interest are the set functions taking values in $\mathbb{R}_+$ or in $\mathbb{R}_+ \cup \{\infty\}$ (but it is convenient sometimes to consider real or even vector-valued set functions).

Definition 1.1.3 A set function $\mu : \mathcal{L} \to \mathbb{R}_+ \cup \{\infty\}$ is additive (precisely, finitely additive) if

$$\mu\left(\bigcup_{i=1}^{n} A_i\right) = \sum_{i=1}^{n} \mu(A_i),$$

for every sequence $(A_i)_{i=1}^n$ of disjoint sets in $\mathcal{L}$; μ is called σ-additive (or countably additive) if

$$\mu\left(\bigcup_{i=1}^{\infty} A_i\right) = \sum_{i=1}^{\infty} \mu(A_i)$$

for every sequence $(A_i)_i$ of pairwise disjoint sets in $\mathcal{L}$ whose union belongs to $\mathcal{L}$.

A positive and additive set function defined on an algebra is called a measure. A measure μ on Ω is finite if $\mu(\Omega) < \infty$ (and σ-finite if $\Omega = \bigcup_{i=1}^{\infty} X_i$, where $X_i \in \mathcal{A}$ and $\mu(X_i) < \infty$ for every i). A probability measure is every measure μ such that $\mu(\Omega) = 1$.

The triplet (Ω, Σ, μ) is referred to as a *measure space* if Σ is a σ-algebra of subsets of Ω and $\mu : \Sigma \to \mathbb{R}_+ \cup \{\infty\}$ is a σ-additive measure. A measure space (Ω, Σ, μ) is called a probability space if $\mu(\Omega) = 1$.

The property of finite additivity is equivalent to the equality

$$\mu(A \cup B) = \mu(A) + \mu(B),$$

for all disjoint sets $A, B \in \mathcal{A}$. As a consequence, every additive set function μ defined on an algebra $\mathcal{A}$ of Ω verifies the property

$$A, B \in \mathcal{A},\ A \subset B \text{ and } \mu(A) < \infty \text{ implies } \mu(B \backslash A) = \mu(B) - \mu(A),$$

which in turn implies the property of *modularity*,

$$\mu(A \cup B) + \mu(A \cap B) = \mu(A) + \mu(B) \quad \text{for all } A, B \in \mathcal{A}$$

and also the property of *subadditivity*,

$$\mu\left(\bigcup_{i=1}^{n} A_i\right) \leq \sum_{i=1}^{n} \mu(A_i)$$

for every finite family $(A_i)_{i=1}^n$ of sets in $\mathcal{A}$.

Notice that every σ-additive measure is *countably subadditive*,

$$\mu\left(\bigcup_{i=1}^{\infty} A_i\right) \leq \sum_{i=1}^{\infty} \mu(A_i)$$

for every sequence $(A_i)_i$ of sets in $\mathcal{A}$ whose union belongs to $\mathcal{A}$.

Example 1.1.4 Let $\mathcal{A}$ be the algebra consisting of all finite subsets of $\mathbb{N}$ and their complements. Then the set function $\mu : \mathcal{A} \to \mathbb{R}_+$ given by $\mu(A) = 0$ if A is finite and $\mu(A) = 1$ otherwise is an example of additive set function that is not σ-additive.

A simple (but important) criterion of σ-additivity is as follows:

Proposition 1.1.5 *An additive set function $\mu : \mathcal{A} \to \mathbb{R}_+$ defined on an algebra is σ-additive if and only if it verifies one of the following equivalent properties:*

(i) (*continuity at zero*) *If $(A_n)_n$ is a decreasing sequence of sets in $\mathcal{A}$ such that $\bigcap_{n=1}^{\infty} A_n = \emptyset$, then*

$$\lim_{n\to\infty} \mu(A_n) = 0;$$

(ii) (*lower continuity or continuity by ascending sequences*) *For every increasing sequence $(A_n)_n$ of sets in $\mathcal{A}$ such that $\bigcup_{n=1}^{\infty} A_n \in \mathcal{A}$, we have*

$$\lim_{n\to\infty} \mu(A_n) = \mu\left(\bigcup_{n=1}^{\infty} A_n\right);$$

(iii) (*upper continuity, or continuity by descending sequences*) *For every decreasing sequence $(A_n)_n$ of sets in $\mathcal{A}$ such that $\bigcap_{n=1}^{\infty} A_n \in \mathcal{A}$, we have*

$$\lim_{n\to\infty} \mu(A_n) = \mu\left(\bigcap_{n=1}^{\infty} A_n\right).$$

See [25, Proposition 1.3.3], for details.

Example 1.1.6 Let Ω be a finite set with N elements. Then the formula $\mu(A) = |A|$ (= cardinality of A) defines a finite σ-additive measure on $\mathcal{P}(\Omega)$, while $\nu(A) = |A|/N$ defines a probability measure.

Another simple example of a probability measure on a set Ω is the *Dirac measure* concentrated at a point $\mathbf{a}$, defined by the formula

$$\delta_{\mathbf{a}} : \mathcal{P}(\Omega) \to [0, 1], \quad \delta_{\mathbf{a}}(A) = \begin{cases} 1 \text{ if } \mathbf{a} \in A \\ 0 \text{ if } \mathbf{a} \notin A. \end{cases}$$

Every positive series $\sum_{n\geq 0} a_n$ gives rise to a measure μ defined on the power set of $\mathbb{N}$ via the formula

$$\mu(A) = \sum\nolimits_{n \in A} a_n;$$

In general, μ is a σ-finite measure; μ is a finite measure if and only if the given series is convergent.

A first step in the construction of Lebesgue measure m on $\mathbb{R}$ is to consider the family $\mathcal{I}$ of all finite unions of bounded intervals of real numbers and to notice that every set A in $\mathcal{I}$ can be represented (nonuniquely) as the union of finitely many disjoint intervals. Assuming that $A = \bigcup_{k=1}^{n} I_k$ is such a decomposition, one defines the Lebesgue measure of A as the sum of the lengths of intervals I_k, that is,

$$m(A) = \sum_{k=1}^{n} \ell(I_k).$$

One can easily show that this definition is independent of the particular decomposition chosen for A and that m is a σ-additive set function on the lattice $\mathcal{I}$. The algebra $\mathcal{A}$ generated by $\mathcal{I}$ consists of all finite unions of intervals (bounded or not) and m can be extended uniquely to a measure on $\mathcal{A}$ by letting $m(A) = \infty$ if A is unbounded.

The next extension of m is based on Carathéodory's method of considering the outer measure attached to m. This is the set function $m^* : \mathcal{P}(\mathbb{R}) \to \mathbb{R}_+ \cup \{\infty\}$ defined by the formula

$$m^*(A) = \inf \left\{ \sum_{i=1}^{\infty} m(A_i) : A \subset \bigcup_{i=1}^{\infty} A_i, \ A_i \in \mathcal{A} \text{ for every } i \right\}.$$

A set $A \subset \mathbb{R}$ is called m^* -measurable (equivalently, Lebesgue measurable) if

$$m^*(H) = m^*(H \cap A) + m^*(H \cap \complement A) \quad \text{for every } H \subset \mathbb{R}.$$

One can prove (see Carothers [32, Chapter 16]), that the family Σ_m of all Lebesgue measurable sets is a σ-algebra that includes $\mathcal{B}(\mathbb{R})$ and the restriction m of m^* to Σ_m is a σ-additive measure that plays the following properties of regularity:

$$m(A) = \inf\{m(U) : U \supset A, \ U \text{ open}\} \text{ for every } A \in \Sigma_m.$$

All subsets $A \subset \mathbb{R}$ with $m^*(A) = 0$ are Lebesgue measurable; they are usually referred to as the *Lebesgue measure zero sets*. Clearly, every subset of a Lebesgue measure zero set is also a measure zero set. An important fact is that every Lebesgue measurable set A can be represented as the union

$$A = B \cup N,$$

between a Borel set B and a Lebesgue measure zero set N. See Carothers [32, Theorem 16.21, p. 283].

Every countable set is a measure zero set. A famous example of a set that is not countable but has measure zero is the Cantor set (and the same occurs for other fractal sets). See [43].

A measure space (Ω, Σ, μ) is called a *complete* measure space if

$$S \subset N, \ N \in \Sigma \text{ and } \mu(N) = 0 \text{ implies } S \in \Sigma.$$

So is the case of the Lebesgue measure space.

Every incomplete measure space (Ω, Σ, μ) admits a *completion* (Ω, Σ', μ'), where Σ' is the σ-algebra of all sets of the form $A \cup S$ with $A \in \Sigma$ and S included in a set $N \in \Sigma$ with $\mu(N) = 0$; μ' is defined by the formula

$$\mu'(A \cup S) = \mu(A).$$

Definition 1.1.7 Given a measure space (Ω, Σ, μ), we say that a property that depends on the variable $x \in \Omega$ holds almost everywhere or a.e. (precisely, μ-almost everywhere or μ-a.e.) if it is true for all values of x except for a set of measure zero.

This definition allows us to speak on functions that are almost everywhere positive (finite almost everywhere, defined almost everywhere, etc.), as well as on sequences of functions converging pointwise almost everywhere.

A basic notion associated with a measurable space (Ω, Σ) is that of measurable function.

Definition 1.1.8 A function $f : \Omega \to \mathbb{R}$ is called measurable (precisely, measurable with respect to the σ-algebra Σ) if $f^{-1}(B) \in \Sigma$, whenever $B \in \mathcal{B}(\mathbb{R})$ (equivalently, see Remark 1.1.2, if $\{x : f(x) < c\} \in \Sigma$ for every $c \in \mathbb{R}$). Their set is denoted by $\mathcal{M}(\Omega, \Sigma)$.

The measurable functions are also called random variables in the context of probability measure spaces.

Every real valued continuous function defined on a Borel subset of $\mathbb{R}^N$ is Borel measurable and every function equal almost everywhere with a Lebesgue-measurable function is also measurable.

If f and g are measurable functions so are $f + g, fg, |f|$, and also $f^+ = \sup\{f, 0\}$, $f^- = \sup\{-f, 0\}$. Every function that is the limit almost everywhere of a sequence $(f_n)_n$ of measurable functions is measurable too. See Hewitt and Stromberg [98, Theorem 11.12, p.153]. These properties imply that $\mathcal{M}(\Omega, \Sigma)$ is a σ-complete vector lattice. See Sect. 1.2 for this terminology.

The *simple functions* i.e., the functions $f : \Omega \to \mathbb{R}$ that assume finitely many values c_i on disjoint sets $A_i \in \Sigma$, are also measurable. Notice that a nonnegative function is measurable if and only if it is the supremum of an increasing sequence of nonnegative simple functions. See Lemma 2.5.5.

By definition, the (μ-) integral of a simple function $\sum_{i=1}^{n} c_i \chi_{A_i}$ is given by the formula

$$\int_{\Omega} \left(\sum_{i=1}^{n} c_i \chi_{A_i}\right) \mathrm{d}\mu(x) = \sum_{i=1}^{n} c_i \mu(A_i)$$

and the integral of a simple function f over a set $A \in \Sigma$ is given by

$$\int_A f \, \mathrm{d}\mu(x) = \int_{\Omega} f \chi_A \, \mathrm{d}\mu(x).$$

The integral is well-defined (i.e., independent of the particular representation of the simple function) due to the additivity of the measure.

Lemma 1.1.9 *The simple functions associated with a measure space* (Ω, Σ, μ) *constitute a vector space denoted* $\mathcal{S}(\Omega, \Sigma)$ *and the integral of simple functions has the following basic properties:*

(i) (Linearity) *If* f, g *are simple functions and* $\alpha, \beta \in \mathbb{R}$, *then*

$$\int_{\Omega} [\alpha f(x) + \beta g(x)] \, \mathrm{d}\mu(x) = \alpha \int_{\Omega} f(x) \, \mathrm{d}\mu(x) + \beta \int_{\Omega} g(x) \, \mathrm{d}\mu(x);$$

(ii) (Positivity) *If* $f \geq 0$ *a.e. then* $\int_{\Omega} f(x) \, \mathrm{d}\mu(x) \geq 0$.

The property of linearity implies the property of *additivity*: if A and B are disjoint sets in Σ, and f is a simple function, then

$$\int_{A \cup B} f(x) \, \mathrm{d}\mu(x) = \int_A f(x) \, \mathrm{d}\mu(x) + \int_B f(x) \, \mathrm{d}\mu(x).$$

The property of positivity implies the property of *monotonicity* i.e.

$$f \leq g \text{ a.e. implies } \int_{\Omega} f(x) \, \mathrm{d}\mu(x) \leq \int_{\Omega} g(x) \, \mathrm{d}\mu(x).$$

Moreover, if $f \in \mathcal{S}(\Omega, \Sigma)$, then the same is true for its absolute value $|f|$ and the property of monotonicity implies the following chain of inequalities:

$$\left| \int_{\Omega} f(x) \, \mathrm{d}\mu(x) \right| \leq \int_{\Omega} |f(x)| \, \mathrm{d}\mu(x) \leq \mu(\Omega) \sup_{x \in \Omega} |f(x)|.$$

Notice that the left-hand inequality is a generalization of the absolute value inequality. As

$$\mu(A) = \int_\Omega \chi_A \, d\mu(x) \quad \text{for all } A \in \Sigma,$$

the integral of simple functions represents a linear extension of the measure μ to a linear and positive operator on $\mathcal{S}(\Omega, \Sigma)$. Remarkably, this extension can be continued one more step to reach the space of all μ-integrable functions. Formally, this is done by completing $\mathcal{S}(\Omega, \Sigma)$ in a manner reminding the completion of normed vector spaces in Functional Analysis.

The natural way to measure the distance between the elements of $\mathcal{S}(\Omega, \Sigma)$ is provided by a substitute of the notion of norm, called the L^1-*seminorm*, that is defined by the formula

$$\|f\|_{L^1} = \int_\Omega |f(x)| \, d\mu(x).$$

Its main properties are very close to those of a norm:

(SN1) $\|f\|_{L^1} \in [0, \infty)$ but it is possible that $\|f\|_{L^1} = 0$ even if f is not identically 0;
(SN2) $\|\alpha f\|_{L^1} = |\alpha| \, \|f\|_{L^1}$;
(SN3) $\|f + g\|_{L^1} \le \|f\|_{L^1} + \|g\|_{L^1}$.

Following the model of $\mathbb{R}$,, one can discuss the problem of convergence of sequences of functions in $\mathcal{S}(\Omega, \Sigma)$.

Definition 1.1.10 A sequence $(f_n)_n$ of simple functions is called fundamental (i.e., fundamental in the mean) if for every $\varepsilon > 0$, there is a number n such that

$$\int_\Omega \left| f_i(x) - f_j(x) \right| d\mu(x) < \varepsilon \quad \text{for all } i, j \ge n.$$

Simple examples show that not every fundamental sequence $(f_n)_n$ of simple functions has a limit in $\mathcal{S}(\Omega, \Sigma)$, i.e., a function $f \in \mathcal{S}(\Omega, \Sigma)$ such that for every $\varepsilon > 0$, there is a number n such that

$$\int_\Omega |f_i(x) - f(x)| \, d\mu(x) < \varepsilon \quad \text{for all } i \ge n.$$

This problem is solved by enlarging the concept of μ-integrability.

Definition 1.1.11 A function $f : \Omega \to \mathbb{R}$ is said to be μ-integrable if there exists a fundamental sequence $(f_n)_n$ of simple functions converging pointwise almost everywhere to f.

By definition, the integral of f with respect to μ is the limit

$$\lim_{n\to\infty} \int_\Omega f_n(x)\,\mathrm{d}\mu(x),$$

whose existence is an easy consequence of the fact that

$$\left|\int_\Omega \left[f_i(x) - f_j(x)\right]\,\mathrm{d}\mu(x)\right| \leq \int_\Omega \left|f_i(x) - f_j(x)\right|\,\mathrm{d}\mu(x) \to 0 \quad \text{as } i, j \to \infty.$$

The integral of f is usually denoted

$$\int_\Omega f(x)\,\mathrm{d}\mu(x) \text{ or } \int_\Omega f\,\mathrm{d}\mu.$$

It is straightforward that the value of the integral is independent of the sequence $(f_n)_n$ involved in its definition.

Since the simple functions are measurable, it follows that the integrable functions are measurable too.

The set $\mathcal{L}^1(\mu)$ of all μ-integrable functions verifies all nice properties mentioned in the case of measurable functions (in particular, it is a vector lattice).

The L^1-seminorm can be extended to $\mathcal{L}^1(\mu)$ by letting

$$\|f\|_{L^1} = \int_\Omega |f|\,\mathrm{d}\mu.$$

The problem with it is the fact that

$$f, g \in \mathcal{L}^1(\mu) \text{ and } f = g \text{ a.e. implies } \|f\|_{L^1} = \|g\|_{L^1},$$

so $\|f - g\|_{L^1} = 0$ does not mean $f = g$. The remedy for this is to replace the vector space $\mathcal{L}^1(\mu)$ with the space $L^1(\mu) = \mathcal{L}^1(\mu)/\sim$, whose elements are the equivalence classes of functions of $\mathcal{L}^1(\mu)$ relative to the relation of equivalence $\sim$, equality almost everywhere. Letting $\hat{f} = \{g : g \sim f\}$ for such classes, it is clear that two classes $\hat{f}$ and $\hat{g}$ either coincide or are disjoint. If we define

$$\hat{f} + \hat{g} = \widehat{f+g}$$
$$\alpha\hat{f} = \widehat{\alpha f}$$
$$\left\|\hat{f}\right\|_{L^1} = \|f\|_{L^1},$$

then $L^1(\mu)$ will become a normed vector space. It is customary to speak of $L^1(\mu)$ as the space of all μ-integrable functions, which is formally incorrect, but convenient. Certainly, it is meant that functions equal almost everywhere are regarded as the same element. See Bogachev [25, p. 250].

One can prove that the space $L^1(\mu)$ is a Banach space (i.e., complete with respect to the norm $\|\cdot\|_{L^1}$) and even a Banach lattice with respect to the order relation

$$f \le g \text{ if and only if } f(x) \le g(x) \text{ a.e.}$$

The construction of the entire scale of spaces $L^p(\mu)$, for $p \in [1, \infty]$, is discussed in detail by many authors. See Bogachev [25], Lieb and Loss [117], and Willem [187].

When μ is the Lebesgue measure on $\mathbb{R}^N$, the space $L^1(\mu)$ is usually denoted $L^1(\mathbb{R}^N)$.

We end our review on measure theory by noticing several criteria of integrability concerning a measure space (Ω, Σ, μ).

Theorem 1.1.12 (The Monotone Convergence Theorem) *Let $(f_n)_n$ be an increasing sequence of nonnegative measurable functions, possibly taking infinite values. Then the function $f = \sup_{n\in\mathbb{N}} f_n$ defined pointwise by the formula*

$$f(x) = \sup_{n\in\mathbb{N}} f_n(x)$$

is measurable and

$$\int_\Omega \sup_{n\in\mathbb{N}} f_n \, \mathrm{d}\mu = \sup_{n\in\mathbb{N}} \int_\Omega f_n \, \mathrm{d}\mu = \lim_{n\to\infty} \int_\Omega f_n \, \mathrm{d}\mu.$$

If, in addition,

$$\sup_{n\in\mathbb{N}} \int_\Omega f_n \, \mathrm{d}\mu < \infty,$$

then $f \in L^1(\mu)$ and

$$\lim_{n\to\infty} \int_\Omega |f - f_n| \, \mathrm{d}\mu = 0.$$

In particular,

$$\lim_{n\to\infty} \int_\Omega f_n \, \mathrm{d}\mu = \int_\Omega f \, \mathrm{d}\mu.$$

An immediate consequence is the fact that

$$\lim_{n\to\infty}\int_{\Omega} f_n \,\mathrm{d}\mu = 0$$

for every sequence $(f_n)_n$ of elements of $L^1(\mu)$, decreasing pointwise to 0.

From the Monotone Convergence Theorem, one can infer the following criterion of integrability.

Theorem 1.1.13 (The Dominated Convergence Theorem) *Let $(f_n)_n$ be a sequence of functions in $L^1(\mu)$ which is almost everywhere convergent to a function $f : \Omega \to \mathbb{R}$. If there exists a function $g \in L^1(\mu)$ such that $|f_n| \leq g$ almost everywhere for all $n \in \mathbb{N}$, then $f \in L^1(\mu)$ and*

$$\lim_{n\to\infty}\int_{\Omega} |f_n - f| \,\mathrm{d}\mu = 0.$$

In particular,

$$\int_{\Omega} f \,\mathrm{d}\mu = \lim_{n\to\infty}\int_{\Omega} f_n \,\mathrm{d}\mu.$$

An important consequence of the Dominated Convergence Theorem is the property of *absolute continuity* of the set functions

$$\nu(A) = \int_A f \,\mathrm{d}\mu, \quad A \in \Sigma,$$

attached to functions $f \in L^1(\mu)$: for every $\varepsilon > 0$, there is $\delta > 0$ such that

$$A \in \Sigma \text{ and } \mu(A) < \delta \text{ implies } \int_A |f| \,\mathrm{d}\mu < \varepsilon.$$

This yields (see Proposition 1.1.5) the σ-additivity of ν and represents an important source of signed measures. A particular case of this construction is the case of the N-dimensional *Gaussian measure,*

$$\gamma_N(A) = \frac{1}{(2\pi)^{N/2}}\int_A e^{-\frac{1}{2}\|\mathbf{x}\|^2}\,\mathrm{d}\mathbf{x}, \quad A \in \mathcal{B}(\mathbb{R}^N).$$

Theorem 1.1.14 (The Comparison Theorem) A function $f : \Omega \to \mathbb{R}$ *is μ-integrable if and only if it is measurable, and there is a μ-integrable function g such that $|f| \leq g$.*

This is a consequence of the Dominated Convergence Theorem.

The notion of measurable function can be extended to the case of functions acting on a pair of measurable spaces. If $(X, \mathcal{A})$ and $(Y, \mathcal{B})$ are two measurable spaces, a function $T : X \to Y$ is called *measurable* (precisely $(\mathcal{A}, \mathcal{B})$-*measurable*) if $T^{-1}(B) \in \mathcal{A}$ for every $B \in \mathcal{B}$. This kind of measurability is related to the following construction of measure spaces $(Y, \mathcal{B}, \nu)$ associated with measure spaces $(X, \mathcal{A}, \mu)$ via the formula

$$\nu(B) = \mu(T^{-1}(B)) \quad \text{for every } B \in \mathcal{B}.$$

The measure ν is called the *push-forward measure* of μ by T and denoted $T\#\mu$.

Theorem 1.1.15 *A function $f : Y \to \mathbb{R}$ is integrable with respect to the measure $\nu = T\#\mu$ if and only if the function $f \circ T$ is integrable with respect to μ and the formula*

$$\int_Y f \, \mathrm{d}\nu = \int_X f \circ T \, \mathrm{d}\mu \tag{1.1}$$

holds. As a consequence, if f is integrable, then

$$\int_B f \, \mathrm{d}\nu = \int_{T^{-1}(B)} f \circ T \, \mathrm{d}\mu \quad \textit{for all } B \in \mathcal{B}.$$

Proof If $f = \chi_B$ for some $B \in \mathcal{B}$, then $f \circ T = \chi_{T^{-1}(B)}$ and the above integral formula reduces to the definition of ν. By linearity, this formula extends to the case of nonnegative simple functions. In the general case of a nonnegative function, one can choose an increasing sequence of nonnegative simple functions $f_n \in \mathcal{S}(\nu)$ that converges to f. Then $f_n \circ T \geq 0$ and $f_n \circ T \nearrow f \circ T$, so the assertion of the theorem follows from the Monotone Convergence Theorem 1.1.12.

In the general case of real-valued functions, one has to apply the preceding reasoning to $|f|$ and to take into account the Comparison Theorem 1.1.14 and the decomposition $f = f^+ - f^-$.

The second part of the theorem follows by replacing f by $f\chi_B$. ∎

1.2 Ordered Banach Spaces

Algebraic structures endowed with order relations play an important role in mathematics. We next review some basic notions and results concerning the ordered Banach spaces. For more, see the textbooks of Aliprantis and Tourky [6], Meyer-Nieberg [129], and Schaefer [162] (as well as the following two papers by Amann [13] and [14]).

An *ordered vector space* is a pair $(E, \leq)$ consisting of a real vector space E and a partial order relation $\leq$ (i.e., a reflexive, antisymmetric, and transitive relation) such that

$$\mathbf{x} \leq \mathbf{y} \text{ implies } \mathbf{x} + \mathbf{z} \leq \mathbf{y} + \mathbf{z} \text{ and } \alpha\mathbf{x} \leq \alpha\mathbf{y} \tag{1.2}$$

for all $\mathbf{x}, \mathbf{y}, \mathbf{z} \in E$ and $\alpha \in [0, \infty)$.

An equivalent for $\mathbf{x} \leq \mathbf{y}$ is $\mathbf{y} \geq \mathbf{x}$. The notation $\mathbf{x} < \mathbf{y}$ (equivalently, $\mathbf{y} > \mathbf{x}$) will mean that $\mathbf{x} \leq \mathbf{y}$ and $\mathbf{x} \neq \mathbf{y}$.

The set $E_+ = \{\mathbf{x} \in E : \mathbf{x} \geq \mathbf{0}\}$ is called the *positive cone* of E. Clearly, $E_+ \cap (-E_+) = \{\mathbf{0}\}$.

Alternatively, the ordered vector spaces can be defined as vector spaces endowed with a pointed convex cone. Recall that a *convex cone* in a real vector space E is any subset C such that

$$C + C \subset C \text{ and } \mathbb{R}_+ C \subset C;$$

equivalently, $\mathbf{x} + \mathbf{y} \in C$ and $\alpha\mathbf{x} \in C$ whenever $\mathbf{x}, \mathbf{y} \in C$ and $\alpha \in \mathbb{R}_+$. The cone C is called *pointed* if $C \cap (-C) = \{\mathbf{0}\}$.

The order relation $\leq$ attached to a pointed convex cone C is defined by the formula

$$\mathbf{x} \leq \mathbf{y} \text{ if and only if } \mathbf{y} - \mathbf{x} \in C$$

and turns E into an ordered vector space such that $E_+ = C$.

The concept of *ordered Banach space* is usually defined as any ordered vector space E endowed with a complete norm (sometimes adding the condition that the positive cone is closed in the norm topology). See [6, p. 85]. However, for the purposes of the this book, we will need a more restrictive concept of ordered Banach space. More precisely, we will consider *only* those ordered Banach spaces E, which verify in addition the following two properties:

(OBS1) the cone E_+ is *generating*, that is, every element of E can be represented as the difference of two elements of E_+;

(OBS2) the norm is *monotone on the positive cone* i.e.,

$$0 \leq \mathbf{x} \leq \mathbf{y} \text{ implies } \|\mathbf{x}\| \leq \|\mathbf{y}\|.$$

Every ordered Banach space E is *Archimedean* i.e., the inequality $nx \leq y$ holds for all $n \in \mathbb{N}$ and some $x, y \in E$ only if $x \leq 0$. See [163, Remark 19(e), p. 254].

An ordered vector space E such that every pair of elements $\mathbf{x}, \mathbf{y}$ admits a supremum $\sup\{\mathbf{x}, \mathbf{y}\}$ and an infimum $\inf\{\mathbf{x}, \mathbf{y}\}$ is called a *vector lattice*. In this case for each $\mathbf{x} \in E$, we can define $\mathbf{x}^+ = \sup\{\mathbf{x}, 0\}$ (the positive part of $\mathbf{x}$), $\mathbf{x}^- = \sup\{-\mathbf{x}, 0\}$ (the negative part of $\mathbf{x}$) and $|\mathbf{x}| = \sup\{-\mathbf{x}, \mathbf{x}\}$ (the modulus of $\mathbf{x}$). We have $\mathbf{x} = \mathbf{x}^+ - \mathbf{x}^-$ and $|\mathbf{x}| = \mathbf{x}^+ + \mathbf{x}^-$. A *Banach lattice* is any real Banach space E, which is at the same time a vector lattice and verifies the compatibility condition

$$|\mathbf{x}| \leq |\mathbf{y}| \text{ implies } \|\mathbf{x}\| \leq \|\mathbf{y}\|.$$

Most (but not all) ordered Banach spaces are actually Banach lattices. The real N-dimensional space $\mathbb{R}^N$ has a natural vector lattice structure associated with the *coordinatewise order*,

$$(x_k)_k \leq (y_k)_k \text{ if and only if } x_k \leq y_k \text{ for all indices } k. \tag{1.3}$$

Each of the norms of index p,

$$\|\mathbf{x}\|_p = \begin{cases} \left(\sum_{k=1}^N |x_k|^p\right)^{1/p} & \text{if } p \in [1, \infty) \\ \sup_{1 \leq k \leq N} |x_k| & \text{if } p = \infty, \end{cases}$$

turns $\mathbb{R}^N$ into a Banach lattice, denoted $\ell^p(N; \mathbb{R})$.

It is customary to denote the N-dimensional Euclidean space $\ell^2(N; \mathbb{R})$ simply by $\mathbb{R}^N$.

The space $\ell^\infty(N; \mathbb{R})$ (i.e. the vector space $\mathbb{R}^N$ endowed with the sup norm and the coordinatewise order) can be identified (algebraically, isometrically and in order) with the Banach lattice $C(\{1, \dots, N\})$, where $\{1, \dots, N\}$ carries the discrete topology.

Other examples of Banach lattices are the spaces of sequences c_0, c and ℓ^p for $1 \leq p \leq \infty$, endowed with the coordinate-wise ordering (1.3), provided that the finite strings are replaced by sequences. The function spaces

$$\begin{aligned}
\mathcal{F}_b(\Omega) &= \{f : \Omega \to \mathbb{R} : \ f \text{ bounded on the set } \Omega\}, \\
C(K) &= \{f : K \to \mathbb{R} : f \text{ continuous on the compact Hausdorff space } K\}, \\
C_b(\Omega) &= \{f : \Omega \to \mathbb{R} : \ f \text{ continuous and bounded on the Hausdorff} \\
&\qquad \text{topological space } \Omega\}, \\
C_c(\Omega) &= \{f : \Omega \to \mathbb{R} : \ f \text{ continuous and having compact support on the} \\
&\qquad \text{locally compact Hausdorff space } \Omega\}, \\
\mathcal{U}C_b(\Omega) &= \{f : \Omega \to \mathbb{R} : \ f \text{ uniformly continuous and bounded on the} \\
&\qquad \text{metric space } \Omega\}, \\
C_0(\Omega) &= \{f : \Omega \to \mathbb{R} : \ f \text{ continuous on the locally compact Hausdorff} \\
&\qquad \text{space } \Omega \text{ and vanishing at infinity}\},
\end{aligned}$$

endowed with the sup-norm $\|f\|_\infty = \sup_{x \in \Omega} |f(x)|$ and the *pointwise order*,

$$f \leq g \text{ if and only if } f(x) \leq g(x) \text{ for all } x \in \Omega,$$

are also examples of Banach lattices.

Other Banach lattices of an utmost interest are the Lebesgue spaces $L^p(\mu) = L^p(\Omega, \Sigma, \mu)$ $(p \in [1, \infty])$, endowed with norm

$$\|f\|_p = \begin{cases} \left(\int_\Omega |f(x)|^p \,\mathrm{d}\mu(x)\right)^{1/p} & \text{if } p \in [1, \infty) \\ \\ \operatorname*{esssup}_{x \in \Omega} |f(x)| & \text{if } p = \infty \end{cases}$$

and the *pointwise ordering modulo null sets* i.e.

$$f \leq g \text{ if and only if } f(x) \leq g(x) \text{ almost everywhere.}$$

A nonempty subset A of an ordered vector space E is *order bounded from above* if there exists an element $\mathbf{z}$ in E (called *upper bound*) such that $\mathbf{x} \leq \mathbf{z}$ for every $\mathbf{x} \in E$. If A admits an order majorant $\mathbf{z}$ that is less than or equal to any other upper bound, then $\mathbf{z}$ is the *least upper bound* or *supremum* of A (abbreviated $\mathbf{z} = \sup A$). Replacing $\leq$ by $\geq$, one obtains respectively the notions of set *bounded from below*, *lower bound* and *infimum* of A (abbreviated $\inf A$). The sets that are order bounded both from above and from below are said to be *order bounded.* So are the open and closed *order intervals*, defined respectively by the formulas

$$(\mathbf{x}, \mathbf{y}) = \{\mathbf{z} : \mathbf{x} < \mathbf{z} < \mathbf{y}\}$$

and

$$[\mathbf{x}, \mathbf{y}] = \{\mathbf{z} : \mathbf{x} \leq \mathbf{z} \leq \mathbf{y}\}.$$

Clearly, every order bounded set can be included in an (open or closed) order interval.

The ordered vector spaces with the property that every nonempty subset that admits an upper bound also admits a supremum are called *order complete.* In their cases, every nonempty subset that admits a lower bound also admits an infimum. Examples of such spaces are $\mathbb{R}^N$, $\mathcal{F}_b(X)$, c_0, ℓ^p and $L^p(\mu)$ for $1 \leq p \leq \infty$. The details can be covered from Aliprantis and Burkinshaw [5]

The vector space $L(\mathbb{R}^N, \mathbb{R}^M)$, of all linear operators $T : \mathbb{R}^N \to \mathbb{R}^M$, is a vector lattice (denoted $L_{\leq}(\mathbb{R}^N, \mathbb{R}^M)$) when endowed with the order relation

$$S \leq T \text{ if and only if } S(\mathbf{x}) \leq T(\mathbf{x}) \text{ for all } \mathbf{x} \geq \mathbf{0}.$$

The corresponding positive cone is

$$L^+(\mathbb{R}^N, \mathbb{R}^M) = \left\{T \in L(\mathbb{R}^N, \mathbb{R}^M) : T(\mathbf{x}) \geq \mathbf{0} \text{ for all } \mathbf{x} \geq \mathbf{0}\right\}$$

and consists of all operators represented by matrices with nonnegative coefficients. As a consequence, if an operator $T \in L_{\leq}(\mathbb{R}^N, \mathbb{R}^M)$ is represented by the matrix $A = (a_{ij})_{i,j}$, then its modulus $|A|$ is represented by the matrix $(|a_{ij}|)_{i,j}$.

The vector space $L(\mathbb{R}^N, \mathbb{R}^M)$ is a Banach space with respect to the operator norm

$$\|T\| = \sup_{\|\mathbf{x}\| \leq 1} \|T(\mathbf{x})\|,$$

no matter which norms we consider on $\mathbb{R}^N$ and $\mathbb{R}^M$. However, the vector lattice $L_{\leq}(\mathbb{R}^N, \mathbb{R}^M)$ becomes a Banach lattice only for *some* choices of the norms on $\mathbb{R}^N$ and $\mathbb{R}^M$, for example, when both these spaces are endowed with the norm of index 1. Indeed, the norm of an operator $T \in L_{\leq}(\ell^1(N; \mathbb{R}), \ell^1(M; \mathbb{R}))$ represented by the matrix $A = (a_{ij})_{i,j}$ is given by the formula

$$\|T\| = \sup_{1 \leq j \leq N} \sum_{i=1}^{M} |a_{ij}|.$$

When $M, N \geq 2$ and $\mathbb{R}^M$ and $\mathbb{R}^N$ are endowed with the Euclidean norms, the vector lattice $L_{\leq}(\mathbb{R}^N, \mathbb{R}^M)$ is no more a Banach lattice with respect to the operator norm.

An important example of ordered Banach space that is not a Banach lattice is $B_{sa}(H)$, the ordered Banach space of all bounded self-adjoint operators acting on a Hilbert space H. This space is endowed with the *Löwner order*,

$$A \leq B \text{ if and only if } \langle A\mathbf{x}, \mathbf{x}\rangle \leq \langle B\mathbf{x}, \mathbf{x}\rangle \quad \text{for all } \mathbf{x} \in H$$

and the operator norm

$$\|A\| = \sup\{\|A(\mathbf{x})\| : \|\mathbf{x}\| \leq 1\}.$$

As is well known, $A \geq 0$ if and only if its spectrum $\sigma(A)$ is included in $[0, \infty)$.

Since $\|A\| = \sup\{|\langle A\mathbf{x}, \mathbf{x}\rangle| : \|\mathbf{x}\| \leq 1\}$ (see Bhatia [24, p. 120]), it follows that $\pm A \leq \|A\| \, \mathrm{I}_H$, where I_H denotes the identity of H. Therefore,

$$\|A\| = \inf\{\|B\| : -B \leq A \leq B\} \quad \text{for all } A \in B_{sa}(H),$$

which shows that

$$0 \leq A \leq B \text{ implies } \|A\| \leq \|B\|.$$

When H is the Euclidean space $\mathbb{R}^N$, the space $B_{sa}(H)$ coincides with the space $\mathrm{Sym}(N, \mathbb{R})$, of all $N \times N$-dimensional symmetric matrices with real coefficients, when endowed with the Löwner ordering and operator norm. Its positive cone,

$$\mathrm{Sym}^+(N,\mathbb{R}) = \left\{A \in \mathrm{Sym}(N,\mathbb{R}) : \langle A\mathbf{x},\mathbf{x}\rangle \geq 0 \text{ for all } \mathbf{x} \in \mathbb{R}^N\right\},$$

consists of all positive semidefinite matrices, while the interior of $\mathrm{Sym}^+(N,\mathbb{R})$ equals the set of all positive definite matrices,

$$\mathrm{Sym}^{++}(N,\mathbb{R}) = \left\{A \in \mathrm{Sym}(N,\mathbb{R}) : \langle A\mathbf{x},\mathbf{x}\rangle > 0 \text{ for all } \mathbf{x} \in \mathbb{R}^N, \mathbf{x} \neq 0\right\}.$$

According to Davies [46], an ordered Banach space E is called *regularly ordered* if

$$\|\mathbf{x}\| = \inf\{\|\mathbf{y}\| : \mathbf{y} \in E,\ -\mathbf{y} \leq \mathbf{x} \leq \mathbf{y}\} \quad \text{for all } \mathbf{x} \in E.$$

This concept offers a large and convenient framework for the theory of ordered Banach spaces.

Lemma 1.2.1 *Every ordered Banach space can be renormed by an equivalent norm to become a regularly ordered Banach space.*

For details, see Namioka [131]. Some other useful properties of ordered Banach spaces are listed below.

Lemma 1.2.2 *Suppose that E is a regularly ordered Banach space. Then:*

(i) *There exists a constant $C > 0$ such that every element $\mathbf{x} \in E$ admits a decomposition of the form $\mathbf{x} = \mathbf{u} - \mathbf{v}$ where $\mathbf{u}, \mathbf{v} \in E_+$ and $\|\mathbf{u}\|, \|\mathbf{v}\| \leq C\|\mathbf{x}\|$.*
(ii) *The dual space of E, E^*, when endowed with the dual cone*

$$E_+^* = \left\{x^* \in E^* : x^*(\mathbf{x}) \geq 0 \textit{ for all } \mathbf{x} \in E_+\right\}$$

is a regularly ordered Banach space.
(iii) *$\mathbf{x} \leq \mathbf{y}$ in E is equivalent to $x^*(\mathbf{x}) \leq x^*(\mathbf{y})$ for all $x^* \in E_+^*$.*
(iv) *$\|\mathbf{x}\| = \sup\left\{x^*(\mathbf{x}) : x^* \in E_+^*,\ \|x^*\| \leq 1\right\}$ for all $\mathbf{x} \in E_+$.*
(v) *If $(\mathbf{x}_n)_n$ is a decreasing sequence of positive elements of E, which converges weakly to 0, then $\|\mathbf{x}_n\| \to 0$.*

The assertion (v) is a generalization of Dini's lemma in Real Analysis; see [43, p. 173].

Proof The assertion (i) follows immediately from Lemma 1.2.1. For (ii), see Davies [46, Lemma 2.4]. The assertion (iii) is an easy consequence of the Hahn–Banach separation theorem; see [137, Theorem B.2.4, p. 417].

The assertion (iv) is also a consequence of the Hahn–Banach separation theorem; see Davies [46, p. 545].

(v) Let K be the convex hull of the set $\{\mathbf{x}_n : n \in \mathbb{N}\}$. According to our hypothesis, 0 is a closure point of K in the weak topology, and thus also in the norm topology, according to a well-known consequence of the Hahn–Banach separation theorem. Hence, for every $\varepsilon > 0$, there exists a convex combination $\sum_{k=1}^{N(\varepsilon)} \lambda_k \mathbf{x}_{n_k}$ of norm less than ε. Put $\widetilde{N}(\varepsilon) = \max\{n_k : k = 1, 2, \ldots, N(\varepsilon)\}$. Then for every $n \geq \widetilde{N}(\varepsilon)$,

$$0 \leq \mathbf{x}_n = \sum_{k=1}^{N(\varepsilon)} \lambda_k \mathbf{x}_n \leq \sum_{k=1}^{N(\varepsilon)} \lambda_k \mathbf{x}_{n_k},$$

which implies that $\|\mathbf{x}_n\| < \varepsilon$. ■

The following definition outlines a class of ordered Banach spaces displaying a strong relationship between the norm and order properties.

Definition 1.2.3 An ordered Banach space E has order continuous norm if every increasing sequence in E, which is order bounded from above is norm convergent.

The order continuity of the norm implies that every increasing (decreasing) sequence $(\mathbf{x}_n)_n$ that is order bounded from above (below) admits a least upper bound (greatest lower bound) $\mathbf{x}$ to which it is norm convergent.

All Banach lattices $\mathbb{R}^n$, c_0, ℓ^p and $L^p(\mu)$ (for $1 \leq p < \infty$) have an order continuous norm. This is not the case for other Banach lattices like c, ℓ^∞ and $C(K)$ (for K a compact Hausdorff space). See the following lemma.

Lemma 1.2.4

(i) *Every ordered Banach space whose closed order intervals are weakly compact has order continuous norm.*
(ii) *A Banach lattice E has order continuous norm if and only if its closed order intervals are weakly compact.*

Proof The assertion (i) follows from Lemma 1.2.2 (v). For (ii), see Meyer-Nieberg [129, Theorem 2.4.2, p. 86]. ■

An example of ordered Banach space that is not a Banach lattice is provided by the space of all self-adjoint compact operators on a Hilbert space, endowed with the cone of positive semidefinite operators. See Glück and Groh [88, Lemma 7.3].

The concept of strictly positive function admits a natural extension in the framework of ordered Banach spaces. Precisely, a positive element $\mathbf{x}$ of an ordered Banach space E is said to be *strictly positive* provided that

$$x^*(\mathbf{x}) > 0 \text{ for every nonzero functional } x^* \in E_+^*.$$

Strictly positive elements exist in every separable ordered Banach space E. For example, choose a sequence $(\mathbf{x}_n)_n$ of elements of E, dense in the closed unit ball and consider their decomposition $\mathbf{x}_n = \mathbf{u}_n - \mathbf{v}_n$ according to Lemma 1.2.2 (i). Then $\mathbf{x} = \sum 2^{-n}(\mathbf{u}_n + \mathbf{v}_n)$ is a strictly positive element since any functional $x^* \in E_+^*$ vanishing at $\mathbf{x}$ will vanish on all elements $\mathbf{x}_n$. By density, this implies that $x^* = 0$. Actually, a stronger result holds, precisely, the set of all strictly positive elements of a separable ordered Banach space is dense into the positive cone. See [163, Theorem 7.6, p. 241].

An example of Banach lattice without strictly positive elements is $C_0(X)$, provided that X is an uncountable set endowed with the discrete metric.

The spaces $C(K)$, $C_b(X)$ and $\mathcal{U}C_b(X)$ (separable or not) admit strictly positive elements with stronger properties, called order units. Recall that an element $\mathbf{u} > 0$ of an ordered Banach space E is called an *order unit* if for each $\mathbf{x} \in E$, there exists a real number $\lambda > 0$ such that $-\lambda\mathbf{u} \leq \mathbf{x} \leq \lambda\mathbf{u}$. Necessarily, an order unit is a strictly positive element.

Another example of ordered Banach space with order unit is $B_{sa}(H)$, the unit being the identity I_H of the Hilbert space H.

Together, these examples illustrate the importance of regularly ordered Banach spaces with an order unit $\mathbf{u}$, whose norms $\|\cdot\|$ are associated with the unit via the formula

$$\|\mathbf{x}\| = \|\mathbf{x}\|_{\mathbf{u}} = \inf\{\lambda > 0 : -\lambda\mathbf{u} \leq \mathbf{x} \leq \lambda\mathbf{u}\}, \quad \mathbf{x} \in E.$$

The natural order unit of the vector lattice $\mathbb{R}^N$ is the vector whose all components equal 1. The norm associated with this unit is the sup-norm.

The following result (due to Gal and Niculescu [81]) explains the ubiquity of operators with values in a $C(K)$ space.

Theorem 1.2.5 *Every ordered Banach space E can be represented as a vector subspace of the space $C(K)$ of continuous real-valued functions on a compact Hausdorff space K via an order preserving linear isometry $\Phi : E \to C(K)$ such that $\Phi(\mathbf{x}) \geq 0$ if and only if $\mathbf{x} \geq 0$. If E has a strictly positive element $\mathbf{e}$, then one can choose K such that $\Phi(\mathbf{e}) = 1$, the unity of $C(K)$.*

Proof According to the Alaoglu theorem, the set $K = \left\{x^* \in E_+^* : \|x^*\| \leq 1\right\}$ is compact relative to the w^* topology. Taking into account the assertions (iii) and (iv) of Lemma 1.2.2, one can easily conclude that E embeds into $C(K)$ via the positive linear isometry

$$\Phi : E \to C(K), \quad (\Phi(\mathbf{x}))(x^*) = x^*(\mathbf{x}).$$

For the second part of Theorem 1.2.5, notice that $\Phi(\mathbf{e})$ is an order unit for $C(K)$. Then the conclusion follows from a classical result due to Kadison [102], stating that every ordered real vector space with an order unit can be represented as a vector subspace of the space of continuous real-valued functions on a compact Hausdorff space via an order-preserving map that carries the order unit to the constant function 1. ■

The positive linear operators acting on ordered Banach spaces are necessarily continuous. The following result appears in a slightly different form in the book of Schaefer and Wolff [163, Theorem 5.5 (ii), p. 228]. Since the way to cover the details is not obvious, we have included here our proof.

Lemma 1.2.6 *Suppose that E and F are ordered Banach spaces and $T : E \to F$ is a linear operator such that $T(\mathbf{x}) \geq 0$ whenever $\mathbf{x} \geq \mathbf{0}$. Then T is continuous.*

Proof According to a remark above, we may assume that both spaces E and F are regularly ordered Banach spaces. A moment's reflection shows that the property of continuity of T is equivalent to the fact that T maps the bounded sequences into bounded sequences. Taking into account Lemma 1.2.2 (i), one can restate this as T maps positive bounded sequences into positive bounded sequences. Thus, if T would be not continuous, then there exists a sequence $(\mathbf{x}_n)_n$ of positive norm-1 elements of E such that $\|T(\mathbf{x}_n)\| \to \infty$. According to the Banach–Steinhaus theorem (the uniform boundedness principle), this implies the existence of a continuous linear functional $y^* \in F^*$ such that $|y^*(T(\mathbf{x}_n))| \to \infty$. Taking into account the assertions (i) and (ii) of Lemma 1.2.2, we may also assume that $y^* \geq 0$. Passing to a subsequence if necessary we arrive at the situation when $y^*(T(\mathbf{x}_n)) \geq 2^n$ for all $n \in \mathbb{N}$. Put $\mathbf{x} = \sum_{n=1}^{\infty} 2^{-n}\mathbf{x}_n$. Then

$$y^*(T(\mathbf{x})) \geq \sum_{k=1}^{n} 2^{-k} y^*(T(\mathbf{x}_k)) \geq n$$

for all $n \in \mathbb{N}$, a contradiction. ■

A nonlinear extension of this result in the context of Banach lattices is provided by Theorem 4.1.10 in Sect. 4.1.

1.3 Some Elements of Convex Analysis

In what follows C is a convex subset of a Banach space E and F is an ordered Banach space.

Given a map $\Phi : C \to F$, we call *subgradient of Φ at a point $x_0 \in C$* every continuous linear operator $T : E \to F$ such that

$$\Phi(x) \geq \Phi(x_0) + T(x - x_0) \quad \text{for all } x \in C. \tag{1.4}$$

The set $\partial_{x_0}\Phi$, of all such subgradients is referred to as the *subdifferential of Φ at x_0* (and the inequality (1.4) is referred to as the *subgradient inequality*). Geometrically, the subgradient inequality asserts that Φ admits a continuous affine minorant $A(x) = \Phi(x_0) + T(x - x_0)$ such that $A(x_0) = \Phi(x_0)$.

The set valued map $x_0 \to \partial_{x_0}\Phi$ represents the *subdifferential* of Φ. The following basic result is due to Valadier [177, p. 68], and Zowe [197, p. 286].

Theorem 1.3.1 *Suppose that C is a convex subset of a Banach space E and F is an order complete Banach space. Then the subdifferential of any continuous convex function $\Phi : C \to F$ at an interior point $x_0 \in C$ is a nonempty convex set.*

Corollary 1.3.2 *Under the hypotheses of Theorem* 1.3.1, *every continuous convex function $\Phi : C \to F$ is the upper envelope of its continuous affine minorants.*

Valadier's argument for the non-emptiness of the subdifferential is based on the existence of directional derivatives

$$\Phi'(x_0; h) = \lim_{t \to 0+} \frac{\Phi(x_0 + th) - \Phi(x_0)}{t},$$

and the fact that a linear operator $T : E \to F$ is a subgradient of Φ at x_0 if and only if $T(h) \le \Phi'(x_0; h)$ for every $h \in E$. He assumed F to be an order complete Banach lattice, a fact relaxed later by Zowe, who used a generalization of the analytical form of the Hahn–Banach Theorem stated below as Theorem 1.3.3.

The converse of Theorem 1.3.1 also works: the only continuous functions $\Phi : C \to F$ with the property that $\partial_{x_0}\Phi \neq \emptyset$ at any interior point x_0 of C are the convex ones. See, e.g., [137, Theorem 4.3.6, p. 149].

Theorem 1.3.3 (The Generalized Hahn–Banach Extension Theorem) *Let Φ be a convex function defined on the real vector space E and taking values in an order complete vector space F. If H is a linear subspace of E and $T_H : H \to F$ is a linear operator satisfying $T_H(x) \le \Phi(x)$ for all $x \in H$, then there exists a linear operator $T : E \to F$, which extends T_H and $T(x) \le \Phi(x)$ for all $x \in E$.*

See Zowe [197, Theorem 2.1, p. 284], for a simple argument.

Corollary 1.3.4 (Topological version of Theorem *1.3.3*) *Let Φ be a convex function defined on the Banach space E and taking values in an order complete Banach space F such that $\Phi(0) = 0$ and Φ is continuous at the origin. If H is a linear subspace of E and $T_H : H \to F$ is a linear operator satisfying the inequality $T_H(x) \le \Phi(x)$ for all $x \in H$, then there exists a continuous linear operator $T : E \to F$ which extends T_H and verifies the inequality*

$$T(x) \le \Phi(x) \quad \text{for all } x \in E. \tag{1.5}$$

Proof Indeed, the existence of a linear operator $T : E \to F$, which extends T_H and verifies the inequality (1.5) follows from Theorem 1.3.3. Since T is linear, its continuity is equivalent to the continuity at the origin. Taking into account that

$$-\Phi(-x) \leq T(x) \leq \Phi(x) \quad \text{for all } x \in E,$$

then for every sequence $x_n \to 0$, we have $-\Phi(-x_n) \leq T(x_n) \leq \Phi(x_n)$ and the fact that $T(x_n) \to 0$ is a consequence of the squeeze theorem. ■

For $H = \{0\}$ and Φ replaced by $\Phi(x + x_0) - \Phi(x_0)$ where x_0 is arbitrarily fixed in E, Corollary 1.3.4 yields the following particular case of Theorem 1.3.1:

Corollary 1.3.5 *Let Φ be a convex function defined on the Banach space E and taking values in an order complete Banach space F. Then $\partial_{x_0}\Phi \neq \emptyset$ at every point x_0 of continuity of Φ.*

Remark 1.3.6 As was noticed in the proof of Corollary 1.3.4, an inequality of the form $T \leq \Phi$ implies the continuity of T on E when $T : E \to F$ is a linear operator and $\Phi : E \to F$ is a convex function continuous at the origin such that $\Phi(0) = 0$. However, the continuity of T also holds by assuming that E is an ordered Banach space and Φ is a monotone increasing function such that $T \leq \Phi$ (the properties of convexity and continuity of Φ being unnecessary in this case).

Indeed,

$$nT(-x) = T(-nx) \leq \Phi(-nx) \leq \Phi(0) \quad \text{for all } n \in \mathbb{N} \text{ and } x \in E_+$$

since Φ is monotone increasing. Taking into account that every ordered Banach space is Archimedean, we infer that $T(-x) \leq 0$ for all $x \in E_+$, that is, T is a positive operator. The continuity of T is now a consequence of Lemma 1.2.6.

The argument of Theorem 1.3.3 can be suitably adapted (see also [141] and [142]) to prove the following variant of it:

Theorem 1.3.7 *Let E be an ordered vector space, F an order complete vector space, $H \subset E$ a vector subspace and $T_1 : H \to F$ a linear operator. If there exists a convex function $\Phi : E_+ \to F$ such that $T_1(h) \leq \Phi(x)$ for all (h, x) in $H \times E_+$ such that $h \leq x$,, then T_1 admits a positive linear extension $T : E \to F$ such that $T|_{E_+} \leq \Phi$.*

It is worth noticing that the hypothesis concerning the order completeness of the codomain F is not only sufficient for the validity of Theorem 1.3.3, but it is also necessary (in the class of ordered vector spaces). See Buskes [30] for a survey on this challenging result due to Bonnice, Silvermann, and To. It is natural to ask whether Theorem 1.3.1 admits a similar companion.

Problem 1.3.8 Suppose that F is an ordered Banach space such that $\partial_{x_0}\Phi \neq \emptyset$ for all continuous convex functions Φ (from an arbitrary Banach space E to F) and all $x_0 \in E$. Is F necessarily an order complete vector space?

A particular class of convex functions is that of sublinear operators. Suppose that E is a vector space, C is a cone included in E, and F is an ordered vector space. A map $P : C \to F$ is called a *sublinear operator* if it is subadditive and positively homogeneous, that is,

$$P(x + y) \leq P(x) + P(y) \text{ and } P(\lambda x) = \lambda P(x)$$

for all $x, y \in C$ and $\lambda \geq 0$. Every sublinear operator is convex and a convex function $\Phi : C \to F$ is sublinear if and only if it is positively homogeneous. In the topological context, the continuity of a sublinear operator $P : C \to F$ is equivalent to its continuity at the origin, which in turn is equivalent to existence of a constant $\lambda \geq 0$ such that

$$\|P(x)\| \leq \lambda \|x\| \quad \text{for all } x \in C.$$

The smallest constant $\lambda = \|P\|$ with this property will be called the *norm* of P.

The monotonicity of the continuous convex functions can be derived via the following consequence of the subgradient inequality.

Lemma 1.3.9 *Suppose that C is an open convex subset of a Banach space E, F is an order complete Banach space and $\Phi : C \to F$ is a continuous convex function. Then for every pair of points x and y in C and every operators $T \in \partial_x\Phi$ and $S \in \partial_y\Phi$, we have*

$$T(y - x) \leq \Phi(y) - \Phi(x) \leq S(y - x).$$

Theorem 1.3.10 (Kutateladze [111]) *Suppose that E is an ordered Banach space, F is an order complete Banach space, and $\Phi : E \to F$ is a continuous convex function. Then the following statements are equivalent:*

(i) *Φ is monotone increasing, i.e., $x \leq y$ in E implies $\Phi(x) \leq \Phi(y)$;*
(ii) *the subdifferential of Φ at any point $x_0 \in E$ consists of positive linear operators;*
(iii) *for each $x_0 \in E$, there exists a positive linear operator $T \in \partial_{x_0}\Phi$.*

Proof (i) $\Rightarrow$ (ii) If $T \in \partial_{x_0}\Phi$, then

$$T(x) \leq \Phi(x) - \Phi(x_0) + T(x_0) \quad \text{for all } x \in E.$$

Since Φ is supposed to be monotone increasing, we have

$$nT(-x) = T(-nx) \leq \Phi(-nx) - \Phi(x_0) + T(x_0) \leq \Phi(0) - \Phi(x_0) + T(x_0)$$

for all $n \in \mathbb{N}$ and $x \in E_+$. Since any ordered Banach space is Archimedean, this implies $T(-x) \leq 0$ and thus T is a positive operator.

The implication (ii) $\Rightarrow$ (iii) is trivial, while the implication (iii) $\Rightarrow$ (i) follows from the left-hand side inequality in Lemma 1.3.9. ■

Things can be considerably simplified under the presence of Gâteaux differentiability.

Suppose that C is an open convex subset of a Banach space E and F is an ordered Banach space. A function $\Phi : C \to F$ is called *Gâteaux differentiable* if for each point $\mathbf{a} \in C$, there is a continuous linear operator $\Phi'(\mathbf{a}) : E \to F$ (called the *Gâteaux derivative* at $\mathbf{a}$) defined by the formula

$$\Phi'(\mathbf{a})(\mathbf{v}) = \lim_{t \to 0^+} \frac{\Phi(\mathbf{a} + t\mathbf{v}) - \Phi(\mathbf{a})}{t} \quad \text{for all } \mathbf{v} \in E.$$

This property of differentiability is weaker than the property of Fréchet differentiability (which occurs when the limit in the above definition exists uniformly for $\mathbf{v}$ in the unit sphere of E).

If a convex function $\Phi : C \to F$ is Gâteaux differentiable, then

$$\Phi(\mathbf{x}) - \Phi(\mathbf{a}) \geq \Phi'(\mathbf{a})(\mathbf{x} - \mathbf{a}) \tag{1.6}$$

for all points $\mathbf{a} \in C$ and $\mathbf{x} \in C$. Moreover, $\partial_{\mathbf{a}}\Phi = \{\Phi'(\mathbf{a})\}$ at all points $\mathbf{a} \in C$. See [137, Section 5.1], for details.

As noticed by Amann [13, Proposition 3.2, p. 184], the Gâteaux differentiability offers a convenient way to recognize the monotone increasing functions acting on ordered Banach spaces: the positivity of their Gâteaux differential.

Lemma 1.3.11 *Suppose that E and F are two ordered Banach spaces, C is a closed convex subset of E with nonempty interior* $\operatorname{int} C$ *and $\Phi : C \to F$ is a function continuous on C and Gâteaux differentiable on* $\operatorname{int} C$. *Then Φ is monotone increasing on C if and only if $\Phi'(\mathbf{a}) \geq 0$ for all* $\mathbf{a} \in \operatorname{int} C$.

Proof The "only if" part follows immediately from the definition of the Gâteaux derivative. For the other implication, notice that the gradient inequality (1.6) shows that Φ is monotone on $\operatorname{int} C$ if $\Phi'(\mathbf{a}) \geq 0$ for all $\mathbf{a} \in \operatorname{int} C$. As regarding the monotonicity on C, that can be established by an approximation argument. Suppose that $\mathbf{x}, \mathbf{y} \in C$ and $\mathbf{x} \leq \mathbf{y}$. For $\mathbf{x}_0 \in \operatorname{int} C$ arbitrarily fixed and $t \in [0, 1)$, both elements $\mathbf{u}_t = \mathbf{x}_0 + t(\mathbf{x} - \mathbf{x}_0)$ and

$\mathbf{v}_t = \mathbf{x}_0 + t(\mathbf{y} - \mathbf{x}_0)$ belong to $\operatorname{int} C$ and $\mathbf{u}_t \leq \mathbf{v}_t$. Moreover, $\mathbf{u}_t \to \mathbf{x}$ and $\mathbf{v}_t \to \mathbf{y}$ as $t \to 1$. Passing to the limit in the inequality $\Phi(\mathbf{u}_t) \leq \Phi(\mathbf{v}_t)$, we conclude that $\Phi(\mathbf{x}) \leq \Phi(\mathbf{y})$. ■

Remark 1.3.12 If the ordered Banach space E has finite dimension, then the statement of Lemma 1.3.11 remains valid by replacing the interior of C by the relative interior of C. See [137, Exercise 6, p. 81].

Numerous examples of monotone increasing convex functions are exhibited in Chaps. 4 and 5. See also [137, Section 5.6].

2 The Choquet Integral

An important fact concerning the integrable functions $f : \Omega \to \mathbb{R}_+$ with respect to a measure space (Ω, Σ, μ) is the following formula:

$$\int_\Omega f \, d\mu = \int_0^\infty \mu\left(\{x \in \Omega : f(x) \geq t\}\right) dt, \tag{2.1}$$

known as the *Layer Cake Representation*. To prove this, notice that the values $f(x)$ of f can be represented as

$$f(x) = \int_0^\infty \chi_{[0, f(x)]}(t) \, dt = \int_0^\infty \chi_{\{y \in \Omega : f(y) \geq t\}}(x) \, dt,$$

which implies that

$$\begin{aligned} \int_\Omega f \, d\mu &= \int_\Omega \left(\int_0^\infty \chi_{\{y \in \Omega : f(y) \geq t\}}(x) \, dt \right) d\mu \\ &= \int_0^\infty \mu\left(\{x \in \Omega : f(x) \geq t\}\right) dt, \end{aligned}$$

the last step being motivated by the Fubini–Tonelli theorem (see Evans and Gariepy [59, Theorem 1, p. 22], or Grigoryan [94, Theorem 2.24, p. 83]).

The Layer Cake Representation can be used to define the integral of a nonnegative function outside the framework of classical measure theory, by noticing that the monotonicity of the set function μ suffices to turn (2.1) into a definition of $\int_\Omega f \, d\mu$ via a Lebesgue integral.

This led to the concept of capacity (a nonnegative and increasing set function that vanishes at the empty set) and also to the development of a number of theories of

S. G. Gal, C. P. Niculescu, *Choquet Integral and Monotone Sublinear Operators*, Frontiers in Mathematics, https://doi.org/10.1007/978-3-031-71701-7_2

integration with respect to a capacity. In this book, we will focus mainly on the Choquet integral and its generalization to the case of vector-valued capacities.

Some historical aspects concerning the theory of capacities are presented in the Comments section, at the end of this chapter.

2.1 Capacities

An important generalization of the notion of measure, with numerous applications to combinatorics, optimization, risk theory, etc., is that of capacity.

Definition 2.1.1 A *capacity* on a set Ω is any set function $\mu : \mathcal{A} \to \mathbb{R}_+ \cup \{\infty\}$ defined on an algebra $\mathcal{A}$ of subsets of Ω that verifies the following two conditions:

$(C1)$ $\mu(\emptyset) = 0$; and
$(C2)$ $\mu(A) \leq \mu(B)$ for all $A, B \in \mathcal{A}$, with $A \subset B$ (monotonicity).

We will refer to the triplet $(\Omega, \mathcal{A}, \mu)$ as a *capacity space*. Except the cases expressly mentioned, the notation (Ω, Σ, μ) will designate a capacity space built on a measurable space (Ω, Σ).

A capacity μ is called *finite* if $\mu(\Omega) < \infty$, *normalized* if $\mu(\Omega) = 1$, and σ*-finite* if Ω is the union of an increasing sequence of sets $A_n \in \mathcal{A}$ such that $\mu(A_n) < \infty$.

To every finite capacity μ, one can attach a new capacity $\overline{\mu}$, the *dual* of μ, which is defined by the formula

$$\overline{\mu}(A) = \mu(\Omega) - \mu(\Omega \setminus A), \quad A \in \mathcal{A}.$$

Notice that $\overline{(\overline{\mu})} = \mu$.

Every finitely additive set function $\mu : \mathcal{A} \to \mathbb{R}_+ \cup \{\infty\}$ is a capacity with the property

$$\mu(A \cup B) + \mu(A \cap B) = \mu(A) + \mu(B) \quad \text{for all } A, B \in \mathcal{A}.$$

In the general context of set functions, this property can be refined.

A set function μ is called *strongly subadditive* (or *submodular,* or *concave*) if

$$\mu(A \cup B) + \mu(A \cap B) \leq \mu(A) + \mu(B) \quad \text{for all } A, B \in \mathcal{A} \tag{2.2}$$

and it is called *strongly superadditive* (or *supermodular,* or 2*-monotone,* or *convex*) if the above inequality works in the other direction, i.e.,

$$\mu(A \cup B) + \mu(A \cap B) \geq \mu(A) + \mu(B) \quad \text{for all } A, B \in \mathcal{A} \tag{2.3}$$

Notice that a finite set function μ defined on an algebra of sets is strongly subadditive (strongly superadditive) if and only if its dual $\overline{\mu}$ is strongly superadditive (strongly subadditive).

Clearly, the strong subadditivity implies *subadditivity*, that is, the inequality

$$\mu(A \cup B) \leq \mu(A) + \mu(B) \quad \text{for all } A, B \in \mathcal{A},$$

while strong superadditivity implies *superadditivity*:

$$\mu(A \cup B) \geq \mu(A) + \mu(B) \quad \text{for all } A, B \in \mathcal{A}.$$

Proposition 2.1.2 *Suppose that* (Ω, Σ) *is a measurable space. Then a set function* $\mu : \Sigma \to \mathbb{R}_+$ *is strongly subadditive if and only if the difference* $\mu(X \cup A) - \mu(X)$ *is a decreasing function of* $X \in \Sigma$*, for every fixed* $A \in \Sigma$.

The proof is immediate.

A normalized strongly subadditive capacity μ defined on an algebra of sets has the property

$$\mu(A) = 0 \text{ implies } \mu(\complement A) = 1. \tag{2.4}$$

Example 2.1.3 Let $f : \mathbb{R} \to \mathbb{R}$ be a continuous functions. With the convention $\sup \emptyset = 0$, the set function $\mu(A) = \sup f(A)$ is strongly subadditive on the finite subsets (or on the compact subsets, or on the bounded subsets) of $\mathbb{R}$. Except for some trivial cases, this is not an additive measure.

Example 2.1.4 The set function $\mu : \mathcal{P}(\mathbb{R}^N) \to \mathbb{N}$ defined by $\mu(\emptyset) = 0$ and $\mu(A) = \dim(\operatorname{span}(A))$, for every nonempty subset A of $\mathbb{R}^N$ is a strongly subadditive capacity.

The next definition formalizes two important notions of continuity for capacities.

Definition 2.1.5 A set function μ is called *lower continuous* (or continuous by ascending sequences) if

$$\lim_{n\to\infty} \mu(A_n) = \mu\left(\bigcup\nolimits_{n=1}^{\infty} A_n\right)$$

for every increasing sequence $(A_n)_n$ of sets in $\mathcal{A}$ such that $\bigcup_{n=1}^{\infty} A_n \in \mathcal{A}$; μ is called *upper continuous* (or continuous by descending sequences) if

$$\lim_{n\to\infty} \mu(A_n) = \mu\left(\bigcap\nolimits_{n=1}^{\infty} A_n\right)$$

for every nonincreasing sequence $(A_n)_n$ of sets in $\mathcal{A}$ such that $\bigcap_{n=1}^{\infty} A_n \in \mathcal{A}$.

If $\mathcal{A}$ is a σ-algebra, then a capacity $\mu : \mathcal{A} \to [0, 1]$ is lower (upper) continuous if and only if its dual $\bar{\mu}$ is upper (lower) continuous.

The upper and lower continuous set functions are generalizations of the countable additive measures. Indeed, if μ is an additive measure defined on a σ-algebra, then the properties of upper/lower continuity are equivalent to countable additivity.

In order to present an important class of lower continuous capacities (with additional properties), we need the following classical result from the theory of convex functions. See [137], for various generalizations.

Lemma 2.1.6 *Let $a < b \leq c < d$ be real numbers such that $a + d = b + c$. Then, every convex function defined on the interval $[a, d]$ verifies the inequality*

$$f(b) + f(c) \leq f(a) + f(d).$$

The inequality works in the opposite direction in the case of concave functions.

Proof According to our hypothesis, there are numbers $r, s \in (0, 1)$ such that $b = (1 - r)a + rd$ and $c = (1 - s)a + sd$. Then $a + d = b + c = (2 - r - s)a + (r + s)d$, whence $(r + s - 1)(d - a) = 0$, i.e., $r + s = 1$. Since f is convex,

$$f(b) + f(c) \leq [(1 - r)f(a) + rf(d)] + [(1 - s)f(a) + sf(d)] = f(a) + f(d).$$

In the case of concave functions, the last inequality should be reversed. ■

Example 2.1.7 (Distorted Measures) There are several standard procedures to attach to a σ-additive measure certain not necessarily additive capacities. So is the case of *distorted measures*,

$$\mu_\varphi(A) = \varphi(\mu(A)),$$

obtained from a σ-additive measure $\mu : \Sigma \to \mathbb{R}_+ \cup \{\infty\}$ by composing it with an increasing and continuous function $\varphi : [0, \mu(\Omega)] \to \mathbb{R}_+ \cup \{\infty\}$ (called distortion), such that $\varphi(0) = 0$. Clearly, μ_φ is a capacity. It is also lower continuous because every increasing sequence $(A_n)_n$ of sets in Σ verifies the equality

$$\bigcup_{n=1}^{\infty} \varphi(A_n) = \varphi\left(\bigcup_{n=1}^{\infty} A_n\right).$$

The property of upper continuity holds in the case of one-to-one distortions, since in that case

$$\bigcap_{n=1}^{\infty} \varphi(A_n) = \varphi\left(\bigcap_{n=1}^{\infty} A_n\right).$$

In the case of concave distortions, the capacity μ_φ is strongly subadditive. So is the case of the capacity m^p, obtained by distorting the Lebesgue measure m, via the function $\varphi(t) = t^p$ with $0 < p \leq 1$. To prove this, consider two sets $A, B \in \Sigma$ with $A \subset B$. Then $\mu(A \cap B) \leq \mu(A) \leq \mu(B) \leq \mu(A \cup B)$ and

$$\frac{\mu(A \cap B) + \mu(A \cup B)}{2} = \frac{\mu(A) + \mu(B)}{2},$$

which imply (due to the fact that φ is concave) that $\varphi(\mu(A \cap B)) + \varphi(\mu(A \cup B)) \leq \varphi(\mu(A)) + \varphi(\mu(B))$. See Lemma 2.1.6.

A similar argument shows that the distorted measure μ_φ is strongly superadditive when φ is convex. For example, this is the case when μ is a probability measure and $\varphi : [0, 1] \to [0, 1]$ is given by $\varphi(x) = x^p$ for $p > 1$.

Another construction to generate capacities with nice regularity condition makes the objective of Theorem 2.2.9.

The set function $H : \mathcal{P}(\mathbb{N}) \to \{0, 1\}$ defined by

$$H(A) = \begin{cases} 0 \text{ if } A \text{ is finite} \\ 1 \text{ if } A \text{ is infinite} \end{cases}$$

is a normalized strongly subadditive capacity, which fails the properties of countable subadditivity, lower continuity, or upper continuity. This capacity plays a role in understanding both the limit superior and the limit inferior of a sequence. See Example 2.8.5 and Remark 2.8.6.

The following simple example of strongly superadditive capacity that is not a distorted measure is due to Chateauneuf. See [37, Example 4, p. 364].

Let $\Omega = \{1, 2, 3, 4\}$ and let $\nu : \mathcal{P}(\Omega) \to [0, 1]$ be the set function defined by:

$$\mu(\{1\}) = \mu(\{3\}) = \frac{1}{5}, \ \nu(\{2\}) = \nu(\{4\}) = \nu(\{2, 4\}) = \frac{1}{6},$$

$$\nu(\Omega) = \frac{1}{10}, \text{ and for any other set } A \subset \Omega, \ \nu(A) = 0.$$

Associated with ν is the strongly superadditive capacity $\mu : \mathcal{P}(\Omega) \to [0, 1]$ defined by

$$\mu(A) = \sum_{B \in \mathcal{P}(\Omega),\ B \subset A} \nu(B).$$

Supposing that μ is a distorted measure, then μ would admit a decomposition $\mu = \varphi \circ P$, for some probability measure P and some nondecreasing function $\varphi : [0, 1] \to [0, 1]$ such that $\varphi(0) = 0$. Since

$$\mu(\{1\}) = \mu(\{3\}) = \frac{1}{5} > \frac{1}{6} = \mu(\{2\}) = \mu(\{4\})$$

we should have $P(\{1\}) > P(\{2\})$ and $P(\{3\}) > P(\{4\})$ since φ is nondecreasing, and hence, it follows that $P(\{1, 3\}) > P(\{2, 4\})$. However, since $\mu(\{1, 3\}) = 2/5 < 1/2 = \mu(\{2, 4\})$ this is a contradiction.

Remark 2.1.8 (Outer and Inner Capacities) Given a capacity space $(\Omega, \mathcal{A}, \mu)$, one can extend μ in at least two ways to a capacity defined on the entire power set of Ω (possibly taking values in $\mathbb{R}_+ \cup \{\infty\}$). This is done by considering the outer capacity associated with μ,

$$\mu^*(E) = \inf\{\mu(A) : A \in \mathcal{A},\ A \supset E\}$$

and the inner capacity associated to μ,

$$\mu_*(E) = \sup\{\mu(A) : A \in \mathcal{A},\ A \subset E\}.$$

Notice that:

(i) μ^* and μ_* are respectively the largest and the smallest capacities that extend μ to $\mathcal{P}(\Omega)$;
(ii) μ^* is finite (subadditive, strongly subadditive) if μ has the respective property. Similarly, μ_* is superadditive (strongly superadditive) if μ has the property under attention.
(iii) If $\mathcal{A}$ is a σ-algebra and μ is strongly subadditive and lower continuous, then μ^* is lower continuous.

For our readers' convenience, we will detail here assertion (iii). Following Ash [16], let us consider an increasing sequence $(E_n)_n$ of subsets of Ω and for $\varepsilon > 0$ arbitrarily fixed let us choose sets $A_n \in \mathcal{A}$ such that $A_n \supset E_n$ and $\mu(A_n) \leq \mu^*(E_n) + \varepsilon/2^n$. Then $B_n = \bigcup_{k=1}^{n} A_k \in \mathcal{A}$ and shall show by induction that

$$\mu(B_n) \leq \mu^*(E_n) + \varepsilon \sum_{k=1}^{n} \frac{1}{2^k}. \tag{2.5}$$

The case $n = 1$ is clear. Suppose now that the inequality (2.5) holds. Then, using the strong subadditivity and monotonicity of μ together with the fact that $E_n \subset B_n \cap A_{n+1}$, we infer that

$$\begin{aligned}\mu(B_{n+1}) &\leq \mu(B_n) + \mu(A_{n+1}) - \mu(B_n \cap A_{n+1}) \\ &\leq \mu^*(E_n) + \varepsilon \sum_{k=1}^{n} \frac{1}{2^k} + \mu^*(E_{n+1}) + \frac{\varepsilon}{2^{n+1}} - \mu^*(E_n) \\ &= \mu^*(E_{n+1}) + \varepsilon \sum_{k=1}^{n+1} \frac{1}{2^k}\end{aligned}$$

and according to the method of mathematical induction, it follows that the inequality (2.5) holds for every natural number $n \geq 1$.

Put $E = \bigcup_{k=1}^{\infty} E_k$ and $B = \bigcup_{k=1}^{\infty} A_k$. Then $B \in \mathcal{A},\ E \subset B$ and

$$\begin{aligned}\mu^*(E) \leq \mu(B) &= \lim_{n\to\infty} \mu(B_n) \\ &\leq \lim_{n\to\infty} \left(\mu^*(E_n) + \varepsilon \sum_{k=1}^{n} \frac{1}{2^k} \right) \\ &\leq \lim_{n\to\infty} \mu^*(E_n) + \varepsilon.\end{aligned}$$

As $\varepsilon > 0$ was arbitrarily fixed and $\lim_{n\to\infty} \mu^*(E_n) \leq \mu^*(E)$, we conclude that $\lim_{n\to\infty} \mu^*(E_n) = \mu^*(E)$.

Another construction of capacities, useful to avoid the sets of infinite capacity, is as follows.

Remark 2.1.9 The *restriction* of a capacity $\mu : \mathcal{A} \to \mathbb{R}_+ \cup \{\infty\}$ to a set $A \in \mathcal{A}$ is the capacity $\mu|_A$ defined on $\mathcal{A} \cap A = \{B : B \in \mathcal{A},\ B \subset A\}$, by the formula

$$\mu|_A(B) = \mu(B).$$

Then $\mu|_A$ inherits all properties of μ, and it is finite if A is a set of finite capacity.

Monotonicity of Higher Order The inclusion–exclusion rule in measure theory asserts that every measure $\mu : \mathcal{A} \to \mathbb{R}_+$ verifies the relation

$$\mu\left(\bigcup_{k=1}^{N} A_n\right) = \sum_{I\neq\emptyset,\ I\subset\{1,\ldots,N\}} (-1)^{|I|+1} \mu\left(\bigcap_{k\in I} A_k\right)$$

for every integer $N \geq 2$ and every finite family $A_1, \ldots, A_N$ of elements of $\mathcal{A}$. Use the mathematical induction.

Accordingly, a set function $\mu : \mathcal{A} \to \mathbb{R}_+$ is said to be *monotone of order* $N \geq 1$ if

$$\mu(A) \geq \sum_{I \neq \emptyset,\ I \subset \{1,\ldots,N\}} (-1)^{|I|+1} \mu\left(\bigcap_{k \in I} A_k\right)$$

for every finite family $A, A_1, \ldots, A_N$ of elements of $\mathcal{A}$ such that $A_i \subset A$ for all i. The condition $(C2)$ in the definition of a capacity represents monotonicity of order $N = 1$. The normalized capacities that are monotone of every order are known as *belief measures*.

An example of belief measure is provided by the set function

$$\mu : \mathcal{B}([0, \infty)) \to [0, 1], \quad \mu(A) = \frac{2}{\sqrt{\pi}} \int_{A^2} e^{-x^2/2} dx,$$

where $A^2 = \{x^2 : x \in A\}$. See Shafer [167], for more information.

2.2 More Examples of Capacities

Sometimes it is important to consider capacities with values in $\mathbb{R}_+ \cup \{\infty\}$. So is the case of outer measures.

Definition 2.2.1 An *outer measure* on Ω is any capacity $\mu : \mathcal{P}(\Omega) \to \mathbb{R}_+ \cup \{\infty\}$ which verifies the property of *countable subadditivity*, that is,

$$\mu\left(\bigcup_{n=1}^{\infty} A_n\right) \leq \sum_{n=1}^{\infty} \mu(A_n)$$

for every sequence $A_1, A_2, A_3, \ldots$ of subsets of Ω.

An important example is the *Lebesgue outer measure* m_N^* on $\mathbb{R}^N$, the N-dimensional generalization of the Lebesgue outer measure m^* on real line. By definition, if A is a subset of $\mathbb{R}^N$, then

$$m_N^*(A) = \inf \sum_{k=1}^{\infty} \operatorname{vol}(C_k)$$

where the infimum is taken over all sequences of open N-dimensional intervals C_k such that $A \subset \bigcup_{k=1}^{\infty} C_k$; recall that an N-dimensional open interval C is any product $I_1 \times \cdots \times I_N$ of N open intervals, and its volume is given by $\operatorname{vol}(C) = \ell(I_1) \times \cdots \times \ell(I_N)$. The next steps of constructing the Lebesgue measure m_N from the Lebesgue outer measure m_N^* are similar to those for $N = 1$, described in Sect. 1.1.

It is worth noticing that the construction of the Lebesgue outer measure m_N^* is not affected if instead of open intervals one considers open balls.

Two other examples of outer measures with profound applications in geometric measure theory are as follows.

Example 2.2.2 (Hausdorff Content) Consider the Euclidean space $\mathbb{R}^N$ of dimension $N \geq 2$ and a real number $d \in (0, N]$. The d-Hausdorff content (or d-Hausdorff capacity or the Hausdorff content of dimension d) of a subset $A \subset \mathbb{R}^N$ is defined by the formula

$$\mathcal{H}_\infty^d(A) = \inf\left\{\sum_{i=1}^{\infty} r_i^d : A \subset \bigcup_{i=1}^{\infty} B_{r_i}(x_i)\right\}, \tag{2.6}$$

where the infimum is taken over all countable coverings of A by open balls; here $B_r(x)$ denotes the open ball centered at x with radius $r > 0$.

The Hausdorff content has the following properties:

(H1) $\mathcal{H}_\infty^d(\emptyset) = 0$;
(H2) if $A \subset B$, then $\mathcal{H}_\infty^d(A) \leq \mathcal{H}_\infty^d(B)$;
(H3) if $A \subset \mathbb{R}^N$, then

$$\mathcal{H}_\infty^d(A) = \inf_{A \subset U \text{ and } U \text{ is open}} \mathcal{H}_\infty^d(U);$$

(H4) if $(K_i)_i$ is a decreasing sequence of compact subsets of $\mathbb{R}^N$, then

$$\mathcal{H}_\infty^d\left(\bigcap_{i=1}^{\infty} K_i\right) = \lim_{i\to\infty} \mathcal{H}_\infty^d(K_i);$$

(H5) if $(A_i)_i$ is any sequence of subsets of $\mathbb{R}^N$, then

$$\mathcal{H}_\infty^d\left(\bigcup_{i=1}^{\infty} A_i\right) \leq \sum_{i=1}^{\infty} \mathcal{H}_\infty^d(A_i).$$

According to (H1), (H2) and (H5), the Hausdorff content is an *outer measure*. The Hausdorff content is neither strongly subadditive nor lower continuous.

A useful variant of the Hausdorff content $\mathcal{H}_\infty^d$ is the dyadic version $\widetilde{\mathcal{H}}_\infty^d$ proposed by Yang and Yuan [189]. This replaces the ball coverings with dyadic cube coverings.

Recall that a dyadic interval is any interval of the form $\left[\frac{p}{2^k}, \frac{p+1}{2^k}\right]$, where $p \in \mathbb{Z}$ and $k \in \mathbb{N}$. A dyadic n-dimensional cube is the product of n dyadic intervals having equal length. For example,

$$\left[\frac{p}{2^k}, \frac{p+1}{2^k}\right] \times \left[\frac{s}{2^k}, \frac{s+1}{2^k}\right] \times \left[\frac{r}{2^k}, \frac{r+1}{2^k}\right]$$

is a dyadic 3-dimensional cube when $p, s, r \in \mathbb{Z}$ and $k \in \mathbb{N}$.

The dyadic version $\widetilde{\mathcal{H}}_\infty^d(A)$ of $\mathcal{H}_\infty^d(A)$ is defined by

$$\widetilde{\mathcal{H}}_\infty^d(A) = \inf\left\{\sum_{i=1}^{\infty} \ell(Q_i)^d : A \subset \operatorname{int}\left(\bigcup_{i=1}^{\infty} Q_i\right)\right\}, \tag{2.7}$$

where Q_i are dyadic cubes with side lengths $\ell(Q_i)$ and int denotes the interior.

Yang and Yuan [189] proved that $\widetilde{\mathcal{H}}_\infty^d$ is an exterior measure verifying the properties of strong subadditivity and lower continuity, as well as the fact that

$$\widetilde{\mathcal{H}}_\infty^d(A) = \inf_{A \subset U \text{ and } U \text{ is open}} \widetilde{\mathcal{H}}_\infty^d(U).$$

Example 2.2.3 (Hausdorff Measure) Let A be a subset of the Euclidean space $\mathbb{R}^N$. For $0 \le d \le N$ and $0 < \rho \le \infty$ denote

$$H_\rho^d(A) = \inf\left\{\sum_{k=1}^{\infty} r_k^d : A \subset \bigcup_{k=1}^{\infty} B_{r_k}(x_k) \text{ and } r_k \le \rho \text{ for all } k\right\},$$

where the infimum is taken over all countable coverings of A by open balls whose radii do not exceed ρ. Since $H_\rho^d(A)$ is nonincreasing in ρ, the limit

$$H^d(A) = \lim_{\rho \to 0+} H_\rho^d(A) = \sup_{\rho > 0} H_\rho^d(A)$$

exists but may be infinite. It is called the *d-dimensional Hausdorff measure* of A. One can easily verify that H^d is a metric outer measure, i.e., an outer measure such that

$$H^d(A \cup B) = H^d(A) + H^d(B)$$

if $d(A, B) = \inf\{\|x - y\| : x \in A,\ y \in B\} > 0$. See [150, Corollary B.13, p. 392].

According to a classical result due to Carathéodory, it follows that the Borel subsets of $\mathbb{R}^N$ are H^d-measurable and H^d is a countably additive measure on $\mathcal{B}(\mathbb{R}^N)$. See Evans and Gariepy [59, Theorem 5, p. 9]. Moreover, H^d is invariant under the isometries of $\mathbb{R}^N$. Since the Lebesgue measure m_N is the unique (Borel) measure invariant to translations that attains 1 on the unit N-dimensional cube, it follows that m_N and H^N differ only by a constant factor. Precisely, $m_N = \frac{\pi^{N/2}}{2^N \Gamma(1+N/2)} H^N$. See [59, Theorem 2, p. 70].

It is worth mentioning that the zero-dimensional Hausdorff measure is the number of points in a set (if the set is finite) or ∞ if the set is infinite. Also, the one-dimensional

Hausdorff measure of a rectifiable curve in $\mathbb{R}^N$ is equal to the length of the curve, and the two-dimensional Hausdorff measure of a Lebesgue-measurable subset of $\mathbb{R}^2$ is proportional to the area of the set. See Mattila [125, p. 56] for details.

Remark 2.2.4 Some authors use other definitions for the Hausdorff measures, by taking coverings with arbitrary sets. For example, if N is a strictly positive integer and $A \in \mathcal{B}(\mathbb{R}^N)$, Evans and Gariepy define $\mathcal{H}^N(A)$ as

$$\inf\left\{\sum_{k=1}^{\infty} \frac{\pi^{d/2}}{\Gamma(1+\delta/2)} \left(\frac{\operatorname{diam} U_k}{2}\right)^N : A \subset \bigcup_{k=1}^{\infty} U_k \text{ and } \operatorname{diam} U_k \leq \rho \text{ for all } k\right\},$$

where the infimum is taken over all countable coverings of A by sets $U_i \subset \mathbb{R}^N$ whose diameters $\operatorname{diam} U_i = \sup\{\|x - y\| : x, y \in U_i\}$ do not exceed ρ. In this case, $H^N(A) = m_N(A)$.

Remark 2.2.5 (Hausdorff Dimension) Let A be an arbitrary subset of $\mathbb{R}^N$. Then, for every $\delta \in (0, 1)$ and $t > s$, we have $H^t_\delta(A) \leq \delta^{t-s} H^s_\delta(A)$, whence, letting $\delta \to 0$, we infer that

$$H^s(A) < \infty \text{ implies } H^t(A) = 0 \text{ for } t > s.$$

This implies the existence of a critical value of s at which $H^s(A)$ jumps from ∞ to 0. This critical value,

$$\dim_H A = \inf\left\{s \geq 0 : H^s(A) = 0\right\},$$

is called the Hausdorff dimension of A. See Falconer [60] for many interesting examples of sets having fractional dimension.

We next describe (following Willem [187, pp. 139–142]) a pretty general technique of constructing capacities on the power set of a metric space M. The starting point is provided by any set function $\mu : \mathcal{K} \to [0, \infty]$ defined on the family $\mathcal{K}$ of compact subsets of M and satisfying the following three conditions:

(Ch1) $\mu(\emptyset) = 0$ and $\mu(A) \leq \mu(B)$ if $A \subset B$;
(Ch2) (regularity) For every $K \in \mathcal{K}$ and every $\alpha > \mu(K)$, there exists an open set U such that $K \subset U$ and $\mu(A) < \alpha$ for all $A \in \mathcal{K}$ included in U;
(Ch3) (strong subadditivity) For every $A, B \in \mathcal{K}$,

$$\mu(A \cup B) + \mu(A \cap B) \leq \mu(A) + \mu(B).$$

For example, these properties are verified by the restriction of the Lebesgue measure m_N to the class of compact subsets of $\mathbb{R}^N$.

The extension of μ to $\mathcal{P}(M)$ is done in two steps.

The first step concerns the extension of μ to the class $\mathcal{O}$ of open subsets U of M via the formula

$$\mu(U) = \sup\{\mu(K) : K \in \mathcal{K},\ K \subset U\}.$$

Lemma 2.2.6 *The extension of μ to the class $\mathcal{O}$ is a monotone and strongly subadditive set function, which verifies the following condition of regularity:*

$$\mu(K) = \inf\{\mu(U) : U \in \mathcal{O},\ U \supset K\} \quad \textit{for every } K \in \mathcal{K}.$$

The proof is immediate except the property of strong subadditivity on the class of open sets that needs the following result.

Claim 2.2.7 Let $A, B \in \mathcal{O}$ and $K \in \mathcal{K}$ be such that $K \subset A \cup B$. Then there exist $K', K'' \in \mathcal{K}$ such that $K' \subset A$, $K'' \subset B$, and $K = K' \cup K''$.

Proof The compact sets $K \setminus A$ and $K \setminus B$ being disjoint, there exist disjoint open sets U and V such that $K \setminus A \subset U$ and $K \setminus B \subset V$. The proof concludes by choosing $K' = K \setminus U$ and $K'' = K \setminus V$. ■

Coming back to the proof of the strong subadditivity of μ on the class of open sets, we have to prove that

$$\mu(A \cup B) + \mu(A \cap B) \leq \mu(A) + \mu(B) \quad \text{for every } A, B \in \mathcal{O}.$$

Due to the monotonicity of μ, the nontrivial case is when $\mu(A \cup B) > 0$. Let us arbitrarily choose two numbers $\alpha > 0$ and $\beta \geq 0$ such that $\alpha < \mu(A \cup B)$ and $\beta < \mu(A \cap B)$. Then there are sets $K, J \in \mathcal{K}$ such that $K \subset A \cup B$, $\alpha < \mu(K)$, $J \subset A \cap B$ and $\beta \leq \mu(J)$. According to the Claim 2.2.7, there exist $K', K'' \in \mathcal{K}$ such that $K' \subset A$, $K'' \subset B$, and $K = K' \cup K''$. Without loss of generality, we may assume that $J \subset K' \cap K''$.

Then, due to the fact that μ is strongly subadditive on compact sets, we have

$$\begin{aligned}\alpha + \beta < \mu(K) + \mu(J) &\leq \mu\left(K' \cup K''\right) + \mu\left(K' \cap K''\right) \leq \mu(K') + \mu(K'') \\ &\leq \mu(A) + \mu(B).\end{aligned}$$

The proof ends by taking the supremum in the left-hand side over α and β.

Now we can extend μ to the entire power set $\mathcal{P}(M)$ via the formula

$$\mu(A) = \inf\{\mu(U) : U \in \mathcal{O},\ U \supset A\} \quad \text{for every } A \in \mathcal{P}(M). \tag{2.8}$$

The following two theorems outline the main features of this extension. The details are straightforward.

Theorem 2.2.8 *For every nonincreasing sequence $(K_n)_n$ of compact sets,*

$$\mu\left(\bigcap_n K_n\right) = \lim_{n\to\infty} \mu(K_n),$$

while for every increasing sequence $(U_n)_n$ of open sets,

$$\mu\left(\bigcup_n U_n\right) = \lim_{n\to\infty} \mu(U_n).$$

Theorem 2.2.9 *The set function μ defined by the formula (2.8) is a lower continuous and countably subadditive capacity on $\mathcal{P}(M)$.*

Choquet [41] (see also Landkof [112, Theorem 2.8, p. 156]) has noticed another striking property of this extension, the fact that all Borel subsets E of M are *capacitable* i.e.,

$$\sup\{\mu(K) : K \subset E,\ K \text{ compact}\} = \inf\{\mu(U) : U \supset E,\ U \text{ open}\}. \tag{2.9}$$

An important example illustrating the above discussion is that of Riesz capacity. The *Riesz capacity* $c_{s,p}$ of a compact set $K \subset \mathbb{R}^N$, $1 < p < \infty$, $0 < sp \leq N$, is defined by

$$c_{s,p}(K) = \inf\left\{\|\varphi\|^p_{L^p(\mathbb{R}^N)} : \varphi * \frac{1}{\|x\|^{N-s}} \geq 1 \text{ on } K\right\},$$

where the infimum is taken over all compactly supported infinitely differentiable functions φ on $\mathbb{R}^N$. The capacity $c_{s,p}$ plays a central role in understanding the nature of Sobolev spaces. See Adams and Hedberg [3].

A short presentation of the Bessel capacity can be found in the Comments section at the end of this chapter.

Potential theory is a big source of capacities, which, roughly speaking, measures the size of a set through an interaction energy. Just to give a glimpse on this vast subject, let us recall that the fundamental solution of Laplace equation $\Delta = 0$ in the space $\mathbb{R}^3$ is

$$K(x, y) = \frac{1}{\|x - y\|}.$$

Given a nonempty compact subset K of $\mathbb{R}^3$, the potential and the energy of a Borel measure μ on K are defined by the formulas

$$U^\mu(x) = \int_{\mathbb{R}^N} K(x, y)\mathrm{d}\mu(y)$$

and

$$\mathcal{E}^{\mu} = \int_{\mathbb{R}^N} U^{\mu}(x)\mathrm{d}\mu(x) = \int_{\mathbb{R}^N}\int_{\mathbb{R}^N} K(x, y)\mathrm{d}\mu(x)\mathrm{d}\mu(y).$$

The *capacity* of the nonempty compact set K is defined as

$$\mathrm{C}(K) = \sup \mu(K),$$

where the supremum is taken over the set of all Borel measures μ on K such that $U^{\mu} \leq 1$ on K.

One can prove that actually there exists a measure μ such that $U^{\mu} = 1$ a.e. on K and $\mu(K) = \mathrm{C}(K)$. Moreover,

$$\mathrm{C}(K) = \frac{1}{\inf \mathcal{E}^{\nu}},$$

where the infimum is taken over all Borel probability measures ν on K. Notice also that $\mathrm{C}(K)$ verifies the hypotheses (Ch1)–(Ch3), so Theorems 2.2.8 and 2.2.9 apply in this case. This is detailed in Landkof [112, Ch. 2]. See also Carleson [31] and Wermer [185].

2.3 Measurable Functions

The Layer Cake Representation can be used (as Choquet [41] did) to define the integral of a nonnegative function with respect to a capacity.

For this, we need to extend the notion of measurability introduced in Sect. 1.1, from the case of σ-algebras, to the case of algebras of sets. For this, consider an arbitrary set Ω endowed with an algebra $\mathcal{A}$ of subsets of Ω.

Definition 2.3.1 A function $f : \Omega \to \mathbb{R}$ is called $\mathcal{A}$-*measurable* (or, simply, measurable) if all sets $\{x : f(x) \in I\}$ belong to $\mathcal{A}$, whenever I is a subinterval of $\mathbb{R}$.

Since $\mathcal{A}$ is an algebra, this notion of measurability is equivalent to *upper measurability*, that asks the membership to $\mathcal{A}$ of all upper level sets $\{x : f(x) \geq t\}$ and $\{x : f(x) > t\}$ for $t \in \mathbb{R}$.

We denote by $\mathcal{M}(\Omega, \mathcal{A})$ the set of all $\mathcal{A}$-measurable functions $f : \Omega \to \mathbb{R}$ and by $\mathcal{M}^{+}(\Omega, \mathcal{A})$ the set of all nonnegative such functions. For technical reasons, in this book, we will be mostly concerned with those measurable functions that are also bounded. We put

$$\mathcal{M}_b(\Omega, \mathcal{A}) = \{f : f : \Omega \to \mathbb{R} \text{ measurable and bounded}\}$$
$$\mathcal{M}_b^{+}(\Omega, \mathcal{A}) = \{f : f \in \mathcal{M}_b(\Omega, \mathcal{A}),\ f \geq 0\}.$$

$\mathcal{M}_b(\Omega, \mathcal{A})$ is a subset of the Banach lattice $\mathcal{F}_b(\Omega)$, of all bounded functions $f : \Omega \to \mathbb{R}$, endowed with the sup norm

$$\|f\| = \sup \{|f(x)| : x \in \Omega\}$$

and the pointwise order

$$f \leq g \text{ if and only if } f(x) \leq g(x) \text{ for all } x \in \Omega.$$

Except the case when $\mathcal{A}$ is a σ-algebra, $\mathcal{M}_b(\Omega, \mathcal{A})$ is not closed under addition or multiplication. However, a special case when the sum of two measurable functions is still a measurable function is addressed in Proposition 2.4.7. Notice also that $\mathcal{M}_b^+(\Omega, \mathcal{A})$ verifies the property

$$f \in \mathcal{M}_b^+(\Omega, \mathcal{A}) \text{ and } \alpha, p \in \mathbb{R}_+ \text{ implies } \alpha f^p \in \mathcal{M}_b^+(\Omega, \mathcal{A})$$

and a similar result holds for $\mathcal{M}^+(\Omega, \mathcal{A})$.

Remark 2.3.2 The set $\mathcal{M}_b(\Omega, \mathcal{A})$ is a a *Stone lattice* of functions in the sense that if $f, g \in \mathcal{M}_b(\Omega, \mathcal{A})$, then $f \vee g$ and $f \wedge g$ also belong to $\mathcal{M}_b(\Omega, \mathcal{A})$ and

$$\alpha + \beta f \in \mathcal{M}_b(\Omega, \mathcal{A})$$

whenever $f \in \mathcal{M}_b(\Omega, \mathcal{A})$ and $\alpha, \beta \in \mathbb{R}$. Recall that

$$(f \vee g)(x) = \max \{f(x), g(x)\} \quad \text{and} \quad (f \wedge g)(x) = \min \{f(x), g(x)\}$$

for all $x \in \Omega$. To prove that $\mathcal{M}_b(\Omega, \mathcal{A})$ is indeed a lattice, use formulas like

$$\{x : (f \vee g)(x) > t\} = \{x : f(x) > t\} \cup \{x : g(x) > t\}$$
$$\{x : (f \vee g)(x) < t\} = \{x : f(x) < t\} \cap \{x : g(x) < t\}.$$

Clearly, $\mathcal{M}(\Omega, \mathcal{A})$ and $\mathcal{M}^+(\Omega, \mathcal{A})$ are also examples of Stone lattices.

Example 2.3.3 Consider the algebra $\mathcal{A}$ of all subsets A of $\mathbb{N}$ such that either A or $\mathbb{N}\backslash A$ is a finite set. Then the functions

$$f(x) = \begin{cases} x & \text{if } x \text{ is odd} \\ x^2 & \text{if } x \text{ is even} \end{cases} \quad \text{and} \quad g(x) = \begin{cases} -x^2 & \text{if } x \text{ is odd} \\ x & \text{if } x \text{ is even} \end{cases}$$

are $\mathcal{A}$-measurable, but their sum and their product fail this property. The details are left to the reader as an exercise.

The situation is much better when algebras are replaced by σ-algebras.

Lemma 2.3.4 *Suppose that Σ is a σ-algebra of subsets of a set Ω. Then the Σ-measurability of a function $f : \Omega \to \mathbb{R}$ is equivalent to each one of the following conditions:*

(i) $\{x : f(x) \geq t\} \in \Sigma$ *for all* $t \in \mathbb{R}$;
(ii) $\{x : f(x) < t\} \in \Sigma$ *for all* $t \in \mathbb{R}$
(iii) $\{x : f(x) \leq t\} \in \Sigma$ *for all* $t \in \mathbb{R}$;
(iv) $\{x : f(x) > t\} \in \Sigma$ *for all* $t \in \mathbb{R}$.

Proof Clearly, (i) $\Leftrightarrow$ (ii) and (iii) $\Leftrightarrow$ (iv). For the other implications notice that

$$\{x : f(x) < t\} = \bigcup\nolimits_{n=1}^{\infty} \{x : f(x) \leq t - 1/n\}$$

and

$$\{x : f(x) > t\} = \bigcup\nolimits_{n=1}^{\infty} \{x : f(x) \geq t + 1/n\}$$

for all $t \in \mathbb{R}$. ■

Lemma 2.3.5 *If Σ is a σ-algebra of subsets of a set Ω, then both spaces $\mathcal{M}(\Omega, \Sigma)$ and $\mathcal{M}_b(\Omega, \Sigma)$ are σ-complete vector lattices, closed under multiplication.*

Proof It suffices to consider the case of $\mathcal{M}(\Omega, \Sigma)$. The fact that

$$f, g \in \mathcal{M}(\Omega, \Sigma) \text{ implies } f + g \in \mathcal{M}(\Omega, \Sigma)$$

is a consequence of the following two formulas:

$$\{x : f(x) > g(x)\} = \bigcup_{r \in \mathbb{Q}} \big(\{x : f(x) > r\} \cap \{x : g(x) < r\}\big),$$

which shows that $\{x : f(x) > g(x)\}$ is a measurable set, and

$$\{x : f(x) + g(x) > \alpha\} = \{x : f(x) > \alpha - g(x)\},$$

which yields the measurability of $f + g$. Hence, according to Remark 2.3.2, it follows that $\mathcal{M}(\Omega, \Sigma)$ is a vector lattice.

As concerns the order σ-completeness of $\mathcal{M}(\Omega, \Sigma)$ it suffices to show that for every increasing sequence $(f_n)_n$ of functions in $\mathcal{M}(\Omega, \Sigma)$, which is bounded from above, the function $f(x) = \sup\{f_n(x) : n \in \mathbb{N}\}$ also belongs to $\mathcal{M}(\Omega, \Sigma)$. This follows from Lemma 2.3.4 and the formula

$$\{x : f(x) > t\} = \bigcup\nolimits_{n=1}^{\infty} \{x : f_n(x) > t\} \quad \text{for all } t \in \mathbb{R}.$$

If $f \in \mathcal{M}(\Omega, \Sigma)$, then also $|f|$ and $f^2 = |f|^2$ are measurable functions. As a consequence, the fact that $\mathcal{M}(\Omega, \Sigma)$ is closed under multiplication follows from the preceding discussion and the formula

$$fg = \frac{(f+g)^2 - (f-g)^2}{4}.$$

■

Corollary 2.3.6 *If Σ is a σ-algebra, then $\mathcal{M}_b(\Omega, \Sigma)$ is an order σ-complete Banach lattice (with respect to the sup norm and pointwise order).*

Remark 2.3.7 (Extended Real-Valued Measurable Functions) Occasionally, we have to consider functions taking infinite values. So is the case of functions defined as pointwise limits. The Definition 2.3.1 of $\mathcal{A}$-measurability can be easily extended to the case of functions $f : \Omega \to \overline{\mathbb{R}}$ by asking that $\{x : f(x) \in I\} \in \mathcal{A}$, whenever I is a subinterval of $\overline{\mathbb{R}}$. Notice that if f is measurable, then the sets $\{x : f(x) = -\infty\}$ and $\{x : f(x) = \infty\}$ are also measurable. When (Ω, Σ) is a measurable space, then the Σ-measurability of a function $f : \Omega \to \overline{\mathbb{R}}$ can be reduced to the fact that

$$f^{-1}((\alpha, \infty]) = \{x : f(x) > \alpha\} \in \Sigma$$

for every $\alpha \in \mathbb{R}$.

2.4 The Choquet Integral Associated with a Capacity

The concept of $\mathcal{A}$-measurability makes possible to associate to each capacity space $(\Omega, \mathcal{A}, \mu)$ and each measurable function $f : \Omega \to \overline{\mathbb{R}}$ an analog of the *survival* (or *reliability*) function,

$$S_f(t) = \mu(\{x : f(x) \geq t\}), \quad t \in \mathbb{R}.$$

The survival function $S_f(t)$ is nonnegative and nonincreasing, so its Lebesgue integral exists. This fact, together with the Layer Cake Representation, led Choquet [41] to extend the concept of Lebesgue integrability by considering capacities rather than measures.

Definition 2.4.1 The Choquet integral of an extended real-valued measurable function $f : \Omega \to \mathbb{R}_+ \cup \{\infty\}$ with respect to a capacity $\mu : \mathcal{A} \to \mathbb{R}_+ \cup \{\infty\}$ is defined by the formula

$$(\mathrm{C}) \int_\Omega f \, \mathrm{d}\mu = \int_0^\infty \mu\left(\{x : f(x) \geq t\}\right) \mathrm{d}t,$$

where the right-hand side integral is a Lebesgue integral.

More generally, the Choquet integral of the function f over a set $A \in \mathcal{A}$ is defined as

$$\begin{aligned}(\mathrm{C}) \int_A f \, \mathrm{d}\mu &= (\mathrm{C}) \int_\Omega f \chi_A \, \mathrm{d}\mu \\ &= \int_0^\infty \mu\left(\{x : f(x) \geq t\} \cap A\right) \mathrm{d}t.\end{aligned}$$

This coincides with the integral of the restriction $f|_A$ (of f to A) with respect to the restricted capacity $\mu|_A$.

Note that in the case of nonnegative, bounded and measurable functions, their survival functions vanish outside a suitable compact interval, so they are Riemann integrable.

The inequality sign $\geq$ in the above formulas can be replaced by $>$.

Lemma 2.4.2 *Under the hypotheses of Definition 2.4.1, we have the equality*

$$\int_0^{+\infty} \mu\left(\{x : f(x) \geq t\} \cap A\right) \mathrm{d}t = \int_0^{+\infty} \mu\left(\{x : f(x) > t\} \cap A\right) \mathrm{d}t$$

for all $A \in \mathcal{A}$ such that $\mu(A) < \infty$.

The condition $\mu(A) < \infty$ can be discarded if μ is a σ-finite and lower continuous capacity.

Proof Let $\varepsilon > 0$ be arbitrarily fixed. Then

$$\begin{aligned}(\mathrm{C}) \int_A f \, \mathrm{d}\mu &= \int_0^\infty \mu\left(\{x : f(x) \geq t\} \cap A\right) \mathrm{d}t \geq \int_0^\infty \mu\left(\{x : f(x) > t\} \cap A\right) \mathrm{d}t \\ &\geq \int_0^{+\infty} \mu\left(\{x : f(x) > t + \varepsilon\} \cap A\right) \mathrm{d}t\end{aligned}$$

$$= \int_{\varepsilon}^{+\infty} \mu\left(\{x : f(x) > t\} \cap A\right) \mathrm{d}t$$

$$\geq \int_{0}^{+\infty} \mu\left(\{x : f(x) > t\} \cap A\right) \mathrm{d}t - \varepsilon\mu(A) = \text{(C)} \int_A f \,\mathrm{d}\mu - \varepsilon\mu(A)$$

and the conclusion follows by letting $\varepsilon \to 0$.

For the second assertion of Lemma 2.4.2, suppose that Ω is the union of an increasing sequence $(A_n)_n$ of measurable sets such that $\mu(A_n) < \infty$. Under the presence of lower continuity,

$$\mu\left(\{x : f(x) \geq t\} \cap A_n\right) \nearrow \mu\left(\{x : f(x) \geq t\}\right)$$

and

$$\mu\left(\{x : f(x) > t\} \cap A_n\right) \nearrow \mu\left(\{x : f(x) > t\}\right).$$

Therefore,

$$\int_0^{\infty} \mu\left(\{x : f(x) \geq t\}\right) \mathrm{d}t = \int_0^{\infty} \mu\left(\{x : f(x) > t\}\right) \mathrm{d}t,$$

according to the classical Monotone Convergence Theorem 1.1.12. ■

According to Definition 2.4.1,

$$\text{(C)} \int_{\Omega} 1_A \,\mathrm{d}\mu = \mu(A) \quad \text{for all } A \in \mathcal{A}$$

and one can easily show that the Choquet integral on $\mathcal{M}^+(\Omega, \mathcal{A})$ has the following property of *invariance to translations*:

$$\text{(C)} \int_{\Omega} (f + \alpha 1_{\Omega}) \,\mathrm{d}\mu = \text{(C)} \int_{\Omega} f \,\mathrm{d}\mu + \alpha \cdot \text{(C)} \int_{\Omega} 1_{\Omega} \,\mathrm{d}\mu \tag{2.10}$$

for every $f \in \mathcal{M}^+(\Omega, \mathcal{A})$ and $\alpha \geq 0$.

We pass now to the problem of defining the Choquet integral in the case of extended real-valued measurable functions. For technical considerations, this will be done *only* for finite capacities.

The starting point is the remark that the Choquet integral admits a unique extension to the set $\mathcal{M}_b(\Omega, \mathcal{A})$ in such a way that the formula (2.10) holds for every $f \in \mathcal{M}_b(\Omega, \mathcal{A})$ and every $\alpha \in \mathbb{R}$. Precisely, this is realized by defining the Choquet integral of a function $f \in \mathcal{M}_b(\Omega, \mathcal{A})$ over a set $A \in \mathcal{A}$ by the formula

$$(\mathrm{C})\int_A f\,\mathrm{d}\mu = (\mathrm{C})\int_A f^+\,\mathrm{d}\mu - (\mathrm{C})\int_A f^-\,\mathrm{d}\overline{\mu}, \tag{2.11}$$

equivalently,

$$(\mathrm{C})\int_A f\,\mathrm{d}\mu = \int_0^{+\infty} \mu\left(\{x : f(x) \geq t\} \cap A\right)\,\mathrm{d}t + \int_{-\infty}^0 \left[\mu\left(\{x : f(x) \geq t\} \cap A\right) - \mu(A)\right]\,\mathrm{d}t. \tag{2.12}$$

If $f \geq 0$, then the last integral in the above formula is 0, so the formula (2.12) really extends Definition 2.4.1 to the whole set $\mathcal{M}_b(\Omega, \mathcal{A})$.

Important It is the presence of the dual capacity in the formula (2.11) (or the presence of the term $-\mu(A)$ in the last integrand in formula (2.12)) that imposes the finiteness of the capacity μ when handling functions with variable sign.

According to Lemma 2.4.2, the right-hand side of the formula (2.12) remains the same if the inequality sign $\geq$ is replaced by $>$.

The property of *invariance to translation* i.e.,

$$(\mathrm{C})\int_\Omega (f + \alpha)\,\mathrm{d}\mu = (\mathrm{C})\int_\Omega f\,\mathrm{d}\mu + \alpha \cdot \mu(\Omega) \tag{2.13}$$

for all $\alpha \in \mathbb{R}$ and all functions $f \in \mathcal{M}_b(\Omega, \mathcal{A})$, follows from the fact that

$$\begin{aligned}(\mathrm{C})\int_\Omega (f + \alpha 1_\Omega)\,\mathrm{d}\mu &= \int_0^{+\infty} \mu\left(\{x : f(x) + \alpha \geq t\}\right)\,\mathrm{d}t \\ &\quad + \int_{-\infty}^0 \left[\mu\left(\{x : f(x) + \alpha \geq t\}\right) - \mu(\Omega)\right]\,\mathrm{d}t \\ &= \int_{-\alpha}^{+\infty} \mu\left(\{x : f(x) \geq s\}\right)\,\mathrm{d}s \\ &\quad + \int_{-\infty}^{-\alpha} \left[\mu\left(\{x : f(x) \geq s\}\right) - \mu(\Omega)\right]\,\mathrm{d}s \\ &= (\mathrm{C})\int_\Omega f\,\mathrm{d}\mu + \alpha \cdot (\mathrm{C})\int_\Omega 1_\Omega\,\mathrm{d}\mu.\end{aligned}$$

Proposition 2.4.3 *The integral given by formula (2.12) is the unique extension of the Choquet integral from $\mathcal{M}_b^+(\Omega, \mathcal{A})$ to $\mathcal{M}_b(\Omega, \mathcal{A})$ which is translation invariant.*

Proof We follow the argument of Marinacci and Montrucchio [124].

Suppose that $I : \mathcal{M}_b(\Omega, \mathcal{A}) \to \mathbb{R}$ is a translation invariant functional that extends the Choquet integral. We have to show that I equals the right-hand side of formula (2.12).

Let $f \in \mathcal{M}_b(\Omega, \mathcal{A})$ be a function such that $\inf_\Omega f = \gamma < 0$. By translation invariance, $I(f - \gamma 1_\Omega) = I(f) - \gamma I(1_\Omega)$. Here 1_Ω denotes the function identically 1 on Ω. Since $f - \gamma 1_\Omega \in \mathcal{M}_b^+(\Omega, \mathcal{A})$, we infer that

$$\begin{aligned}
I(f) = I(f - \gamma 1_\Omega) + \gamma I(1_\Omega) &= \text{(C)} \int_\Omega (f - \gamma 1_\Omega)\, \mathrm{d}\mu + \gamma\mu(\Omega) \\
&= \int_0^{+\infty} \mu(\{x : f(x) - \gamma \geq t\})\, \mathrm{d}t + \gamma\mu(\Omega) \\
&= \int_\gamma^{+\infty} \mu(\{x : f(x) \geq t\})\, \mathrm{d}t + \gamma\mu(\Omega) \\
&= \int_\gamma^0 \mu(\{x : f(x) \geq t\})\, \mathrm{d}t \\
&\quad + \int_0^{+\infty} \mu(\{x : f(x) \geq t\})\, \mathrm{d}t - \int_\gamma^0 \mu(\Omega)\, \mathrm{d}t, \\
&= \int_0^{+\infty} \mu(\{x : f(x) \geq t\})\, \mathrm{d}t + \int_\gamma^0 (\mu(\{x : f(x) \geq t\}) - \mu(\Omega))\, \mathrm{d}t
\end{aligned}$$

whence, by taking into account that $\mu(\{x \in \Omega : f(x) \geq t\}) - \mu(\Omega) = 0$ for $t \leq \gamma$, we conclude that

$$I(f) = \int_0^{+\infty} \mu(\{x : f(x) \geq t\})\, \mathrm{d}t + \int_{-\infty}^0 [\mu(\{x : f(x) \geq t\}) - \mu(\Omega)]\, \mathrm{d}t.$$

The proof is complete. ■

Remark 2.4.4 (Extension of Choquet Integral for Extended Real-Valued Functions) When $\mu(\Omega) < \infty$, the integral of a measurable function $f : \Omega \to \overline{\mathbb{R}}$ can be defined by the formula (2.12), understood as a sum of two Lebesgue integrals,

$$\begin{aligned}
\text{(C)} \int_\Omega f\, \mathrm{d}\mu = \int_0^{+\infty} & \mu(\{x : f(x) \geq t\})\, \mathrm{d}t \\
&+ \int_{-\infty}^0 [\mu(\{x : f(x) \geq t\}) - \mu(\Omega)]\, \mathrm{d}t.
\end{aligned}$$

Equivalently, the integral of f can be defined as a sum of two improper Riemann integrals,

$$(C)\int_{\Omega} f\,d\mu = \lim_{\alpha,\beta\to\infty}\left\{\int_0^{\alpha} \mu\left(\{x : f(x) \geq t\}\right) dt + \int_{-\beta}^{0} \left[\mu\left(\{x : f(x) \geq t\}\right) - \mu(\Omega)\right] dt\right\}.$$

The Choquet integrability of a measurable function $f : \Omega \to \overline{\mathbb{R}}$ with respect to a capacity $\mu : \mathcal{A} \to \mathbb{R}_+ \cup \{\infty\}$ can be defined in terms of Lebesgue integral by asking that

$$(C)\int_{\Omega} |f|\,d\mu = \int_0^{\infty} \mu\left(\{x : |f(x)| \geq t\}\right) dt < \infty.$$

Notice that this approach leads to the Lebesgue integral when $\mathcal{A}$ is a σ-algebra and μ is σ-additive.

However, in the general case, this concept of Choquet integrability **does not imply** the existence of the Choquet integral of the function f unless $\mu(\Omega) < \infty$. More information on this matter is offered by Theorem 2.6.4 in Sect. 2.6. See also Chap. 3, devoted to a discussion about the analogs of L^p spaces.

Some easy (but important) properties of the Choquet integral are summarized as follows:

Proposition 2.4.5 *If $\mu : \mathcal{A} \to \mathbb{R}_+$ is a capacity, $f, g \in \mathcal{M}_b(\Omega, \mathcal{A})$ and $A, B \in \mathcal{A}$, then the associated Choquet integral over $A \in \mathcal{A}$ is a functional on the space $\mathcal{M}_b(\Omega, \mathcal{A})$ such that:*

$$f \geq 0 \text{ implies } (C)\int_A f\,d\mu \geq 0 \quad (\textit{positivity})$$

$$f \leq g \text{ implies } (C)\int_A f\,d\mu \leq (C)\int_A g\,d\mu \quad (\textit{monotonicity})$$

$$A \subset B \text{ and } f \geq 0 \text{ implies } (C)\int_A f\,d\mu \leq (C)\int_B f\,d\mu$$

$$(C)\int_A \alpha f\,d\mu = \alpha\left((C)\int_A f\,d\mu\right) \quad \text{for } \alpha \geq 0 \ (\textit{positive homogeneity})$$

$$(C)\int_A \alpha\,d\mu = \alpha\mu(A) \quad \text{for } \alpha \in \mathbb{R} \ (\textit{calibration})$$

$$(C)\int_A -f\,d\mu = -(C)\int_A f\,d\overline{\mu} \quad (\textit{asymmetry})$$

and

$$\left|(C)\int_A f\,d\mu - (C)\int_A g\,d\mu\right| \leq \mu(A)\,\|f-g\|_{\infty} \quad (\textit{Lipschitz continuity}).$$

Here $\overline{\mu}$ denotes the capacity dual to μ and $\|\cdot\|_\infty$ denotes the sup norm.

Proof For the property of monotonicity, notice that we may restrict to the case where $0 \leq f \leq g$, due to the translation invariance of Choquet integral; see (2.13). In this case, $\{x : f(x) \geq t\} \subset \{x : g(x) \geq t\}$ for every $t \geq 0$, and it remains to take into account that μ is a monotone set function. The property of calibration is also a consequence of the property of translation invariance.

The property of asymmetry is an immediate consequence of the formula

$$(\mathrm{C})\int_A f\,\mathrm{d}\mu = (\mathrm{C})\int_A f^+\,\mathrm{d}\mu - (\mathrm{C})\int_A f^-\,\mathrm{d}\overline{\mu}, \tag{2.14}$$

that formulates the Choquet integral of a function $f \in \mathcal{M}_b(\Omega, \mathcal{A})$ as a combination of the integrals of f^+ and f^-. Indeed,

$$\begin{aligned} -(\mathrm{C})\int_A f^-\mathrm{d}\overline{\mu} &= -\int_0^{+\infty} \overline{\mu}\left(\{x \in A : -f(x) \geq t\}\right)\,\mathrm{d}t \\ &= -\int_0^{+\infty} \left[\mu(A) - \mu\left(A \setminus \{x \in A : -f(x) \geq t\}\right)\right]\mathrm{d}t \\ &= -\int_0^{+\infty} \left[\mu(A) - \mu\left(\{x \in A : f(x) \geq -t\}\right)\right]\mathrm{d}t \\ &= \int_{-\infty}^{0} \left[\mu\left(\{x \in A : f(x) \geq t\}\right) - \mu(A)\right]\mathrm{d}t. \end{aligned}$$

As concerns the Lipschitz continuity, notice that $f \leq g + \|f - g\|_\infty$, which implies (due to the properties of monotonicity and translation invariance) that

$$(C)\int_A f\,\mathrm{d}\mu \leq (C)\int_A g\,\mathrm{d}\mu + (C)\int_A \|f - g\|_\infty\,\mathrm{d}\mu = (C)\int_A g\,\mathrm{d}\mu + \mu(A)\,\|f - g\|_\infty.$$

The proof ends by reversing the roles of f and g. ∎

Proposition 2.4.6 *If μ is a lower continuous capacity, then the Choquet integral is lower continuous in the sense that*

$$\lim_{n\to\infty}\left((C)\int_A f_n\,\mathrm{d}\mu\right) = (C)\int_A f\,\mathrm{d}\mu,$$

whenever $(f_n)_n$ is an increasing sequence of functions in $\mathcal{M}_b(\Omega, \mathcal{A})$ that converges pointwise to a function $f \in \mathcal{M}_b(\Omega, \mathcal{A})$.

This is a consequence of the Monotone Convergence Theorem 1.1.12.

It was already noticed that the set $\mathcal{M}_b(\Omega, \mathcal{A})$ is not closed under addition but $\alpha + \beta f \in \mathcal{M}_b(\Omega, \mathcal{A})$ whenever $f \in \mathcal{M}_b(\Omega, \mathcal{A})$ and $\alpha, \beta \in \mathbb{R}$. Actually, more is true.

Proposition 2.4.7 *The sum of two functions $f, g \in \mathcal{M}_b(\Omega, \mathcal{A})$ is still in the space $\mathcal{M}_b(\Omega, \mathcal{A})$ if one of the functions is simple. In particular, the subset $\mathcal{S}(\Omega, \mathcal{A})$ of all simple functions is a vector space (and even an algebra).*

Proof By translation invariance, we may assume that both functions f and g are nonnegative. Without loss of generality, we may also assume that g is a simple function.

The next step is to remark that $f + \alpha\chi_A \in \mathcal{M}_b^+(\Omega, \mathcal{A})$, whenever $A \in \mathcal{A}$ and $\alpha \geq 0$. Indeed,

$$\{x : f(x) + \alpha\chi_A(x) \geq t\} = \left(\{x : f(x) \geq t\} \cap \complement A\right) \cup (\{x : f(x) \geq t - \alpha\} \cap A) \in \mathcal{A}$$

for every $t \geq 0$. Now the proof can be easily completed by induction. ■

Some simple examples illustrating the computation of Choquet integrals are as follows.

Example 2.4.8 (Infimum and Supremum as Choquet Integrals) Suppose that $\mathcal{A}$ is an algebra of subsets of an arbitrary set Ω and for each $A \in \mathcal{A}$ consider the capacity c_A defined by the formula

$$c_A(B) = \begin{cases} 1 \text{ if } A \text{ is included in } B \\ 0 \text{ if } A \text{ is not included in } B. \end{cases}$$

Then for each $f \in \mathcal{M}_b^+(\Omega, \mathcal{A})$, we have $A \subset \{x \in \Omega : f(x) \geq t\}$ if and only if $t \leq \inf\{f(x) : x \in A\}$. As a consequence,

$$(\mathrm{C}) \int_\Omega f \, \mathrm{d}c_A = \int_0^\infty \chi_{(-\infty, \inf\{f(x):x \in A\}]}(t) \, \mathrm{d}t = \inf\{f(x) : x \in A\}.$$

The translation invariance of the Choquet integral (see formula (2.13)), allows us to conclude that this computation also works when f is real-valued. The representation of supremum as a Choquet integral is provided by the capacity dual to c_A,

$$(\mathrm{C}) \int_\Omega f \, \mathrm{d}\bar{c}_A = -(\mathrm{C}) \int_\Omega (-f) \, \mathrm{d}c_A = \sup\{f(x) : x \in A\}.$$

See the asymmetry formula included in Proposition 2.4.5.

Our next example concerns the important case of simple functions. Recall that a function f in $\mathcal{M}_b(\Omega, \mathcal{A})$ is called *simple* if it takes only finitely many values. Each simple function f admits a unique representation $f = \sum_{i=1}^{n} \alpha_i \chi_{A_i}$, where the sets $A_i \in \mathcal{A}$ ($i \in \{1, \ldots, n\}$) constitute a partition of Ω and $\alpha_1 > \cdots > \alpha_n$.

Example 2.4.9 Consider the case of a simple function $f = \sum_{i=1}^{n} \alpha_i \chi_{A_i} \in \mathcal{M}_b(\Omega, \mathcal{A})$, where the sets $A_i \in \mathcal{A}$ ($i \in \{1, \ldots, n\}$) are pairwise disjoint and $\alpha_1 \geq \cdots \geq \alpha_n$. Then the Choquet integral of f with respect to a capacity $\mu : \mathcal{A} \to \mathbb{R}_+$ is given by the formula

$$
\begin{aligned}
(\mathrm{C}) \int_\Omega f \, \mathrm{d}\mu &= \sum_{i=1}^{n} (\alpha_i - \alpha_{i+1}) \mu \left(\bigcup\nolimits_{j=1}^{i} A_j \right) \\
&= \sum_{i=1}^{n} \alpha_i \left(\mu \left(\bigcup\nolimits_{j=0}^{i} A_j \right) - \mu \left(\bigcup\nolimits_{j=0}^{i-1} A_j \right) \right),
\end{aligned}
$$

with the convention that $\alpha_{n+1} = 0$ and $A_0 = \emptyset$. In the case of additive capacities, this formula becomes

$$
(\mathrm{C}) \int_\Omega f \, \mathrm{d}\mu = \sum_{i=1}^{n} \alpha_i \mu(A_i).
$$

Indeed, taking into account the property of invariance to translation, we may restrict ourselves to the case of nonnegative functions, i.e., to the case where $\alpha_n \geq 0$. We have to prove the first equality, the second being a rearrangement of the first one.

If $t > \alpha_1$, then $\{x : f(x) \geq t\} = \emptyset$, while for $t \in (\alpha_{i+1}, \alpha_i]$ ($i \in \{1, \ldots, n\}$) we have $\{x : f(x) \geq t\} = \bigcup_{j=1}^{i} A_j$. Therefore

$$
\mu \left(\{x : f(x) \geq t\} \right) = \sum_{i=1}^{n} \mu \left(\bigcup\nolimits_{j=1}^{i} A_j \right) \chi_{(\alpha_{i+1}, \alpha_i]}(t)
$$

for all $t \geq 0$, whence

$$
\begin{aligned}
(\mathrm{C}) \int_\Omega f \, \mathrm{d}\mu &= \int_0^{+\infty} \mu \left(\{x : f(x) \geq t\} \right) \mathrm{d}t \\
&= \sum_{i=1}^{n} \mu \left(\bigcup\nolimits_{j=1}^{i} A_j \right) \int_0^{+\infty} \chi_{(\alpha_{i+1}, \alpha_i]}(t) \, \mathrm{d}t \\
&= \sum_{i=1}^{n} (\alpha_i - \alpha_{i+1}) \mu \left(\bigcup\nolimits_{j=1}^{i} A_j \right).
\end{aligned}
$$

Remark 2.4.10 According to Example 2.4.9, if $\Omega = \{1, \ldots, n\}$ is a finite set, then the Choquet integral of any function $f : \Omega \to \mathbb{R}_+$ with respect to a capacity $\mu : \mathcal{P}(\Omega) \to \mathbb{R}_+$ is given by the formula

$$(\mathrm{C})\int_{\Omega} f\,\mathrm{d}\mu=\sum_{i=1}^{n}[f(\sigma(i))-f(\sigma(i-1))]\,\mu(\{\sigma(i),\ldots,\sigma(n)\}), \tag{2.15}$$

where σ is a permutation of Ω such that $f(\sigma(1)) \leq \cdots \leq f(\sigma(n))$ and $f(\sigma(0)) = 0$. When μ is an additive capacity, then the Choquet integral of every function $f : \Omega \to \mathbb{R}$ is given by the formula

$$(\mathrm{C})\int_{\Omega} f\,\mathrm{d}\mu=\sum_{i=1}^{n} f(i)\mu(\{i\}).$$

We will revisit the problem of computing the Choquet integrals in finite terms in Sect. 2.7.

Remark 2.4.11 (Capacities Defined by Densities) Suppose that (Ω, Σ) is a measurable space and $\mu : \Sigma \to \mathbb{R}_+$ is a finite capacity. Then to every Choquet integrable function $f : \Omega \to \mathbb{R}_+ \cup \{\infty\}$, one can attach a new capacity defined by the formula

$$\upsilon(A)=(\mathrm{C})\int_{A} f\,\mathrm{d}\mu,$$

called the *capacity of density* f associated with μ. According to [116], μ and ν verify the following property of absolute continuity: for every $\varepsilon > 0$, there is $\delta > 0$ such that $\nu(A) < \varepsilon$ for all sets $A \in \Sigma$ with $\mu(A) < \delta$.

A very general Radon–Nikodym theorem in this context (including the old result of Graf [91]) can be found in a paper by Ouyang and Li [148].

Also, notice that

$$(\mathrm{C})\int_{A\cup B} f\,\mathrm{d}\mu \leq (\mathrm{C})\int_{A} f\,\mathrm{d}\mu + (\mathrm{C})\int_{B} f\,\mathrm{d}\mu$$

for all $A, B \in \Sigma$, provided that μ is subadditive and $f \geq 0$.

2.5 The Comonotonic Additivity of the Choquet Integral

The Choquet integral with respect to a nonadditive capacity is not additive. Precisely, even if the sum of two functions $f, g \in \mathcal{M}_b(\Omega, \mathcal{A})$ belongs to $\mathcal{M}_b(\Omega, \mathcal{A})$, it might be possible that the equality

$$(C)\int_{\Omega}(f+g)\,\mathrm{d}\mu=(C)\int_{A} f\,\mathrm{d}\mu+(C)\int_{A} g\,\mathrm{d}\mu \tag{2.16}$$

fails. A simple way to check this is to consider the case of two arbitrary disjoint sets $A, B \in \mathcal{A}$ and to notice that the equality

$$(C)\int_\Omega (\chi_A + \chi_B)\,d\mu = (C)\int_\Omega \chi_A\,d\mu + (C)\int_\Omega \chi_B\,d\mu$$

is equivalent to the condition $\mu(A \cup B) = \mu(A) + \mu(B)$, of additivity of μ. The converse also holds: the additivity of the capacity μ implies the additivity of the associated Choquet integral (the equality (2.16), under the requirement that f, g and $f + g$ belong to $\mathcal{M}_b(\Omega, \mathcal{A})$. See Remark 2.6.2.

On the other hand, the property of translation invariance (2.13) provides a case when the additivity holds for every capacity. A much more general result was obtained by Dellacherie [49], using the notion of comonotonicity.

Definition 2.5.1 Two functions $f, g : \Omega \to \mathbb{R}$ are comonotonic if

$$(f(x) - f(y)) \cdot (g(x) - g(y)) \geq 0, \text{ for all } x, y \in \Omega.$$

Every function is comonotonic with any constant function and, if Ω is an ordered set, any two increasing (respectively decreasing) functions are comonotonic. In addition, every positive function is comonotonic with all its powers of positive exponent.

Remark 2.5.2 An important instance of comonotonicity is offered by the pair of functions

$$(f - t)^+ \quad \text{and} \quad f \wedge t,$$

where $f \in \mathcal{M}_b(\Omega, \mathcal{A})$ and $t \in \mathbb{R}$. Notice that $f = (f - t)^+ + f \wedge t$. The case $t = 0$, represents the comonotonicity of the functions f^+ and $-f^-$. See Marinacci and Montrucchio [124, Lemma 3.3].

Comonotonicity is a way to express that two functions have a "similar pattern" of variation. However, this should be understood metaphorically. See the case of the functions

$$f(x) = \begin{cases} -x \text{ if } x \in [-1, 0] \\ 5x \text{ if } x \in [0, 1] \end{cases} \text{ and } g(x) = \begin{cases} -x \text{ if } x \in [-1, 0] \\ 2x \text{ if } x \in [0, 1], \end{cases}$$

whose variation seems similar, but which are not comonotonic because

$$(f(-1/2) - f(1/5))\,(g(-1/2) - g(1/5)) < 0.$$

The next characterization of comonotonicity is due to Denneberg. See [51, Proposition 4.5, p. 54].

Lemma 2.5.3 *Let $f, g : \Omega \to \mathbb{R}$. Then the following assertions are equivalent:*

(i) *f and g are comonotonic;*
(ii) *There is no pair of points $x, y \in \Omega$ such that $f(x) < f(y)$ and $g(x) > g(y)$;*
(iii) *The family of all sets of the form $\{x : f(x) \geq t\}$ or $\{x : g(x) \geq t\}$ with $t \in \mathbb{R}$ is a chain in the sense that for each pair of these sets, one is a subset of the other;*
(iv) *There exist continuous and increasing functions $u, v : \mathbb{R} \to \mathbb{R}$ such that $u + v = id_{\mathbb{R}}$ and*

$$f = u(f + g) \text{ and } g = v(f + g).$$

Proof The equivalences (i) $\Leftrightarrow$ (ii) and (ii) $\Leftrightarrow$ (iii) are straightforward. Also clear is the implication (iv) $\Rightarrow$ (iii).

For (i) $\Rightarrow$ (iv), notice first that every element $z \in (f + g)(\Omega)$ admits a unique representation of the form

$$z = x + y \quad \text{with } z = (f + g)(\omega), \ x = f(\omega) \text{ and } y = g(\omega)$$

for some $\omega \in \Omega$; for this, reason by reductio ad absurdum. We define the functions u and v on $(f + g)(\Omega)$ by

$$u(z) = x \text{ and } v(z) = y.$$

This assures that $u(z) + v(z) = z$ for all $z \in (f + g)(\Omega)$.

The functions u and v are both increasing. Indeed, if $z_1, z_2 \in (f + g)(\Omega)$ and $z_1 < z_2$, then the above discussion yields the existence of two elements ω_1 and ω_2 in Ω such that $f(\omega_1) + g(\omega_1) = z_1 < z_2 = f(\omega_2) + g(\omega_2)$. Or, in this case,

$$f(\omega_1) - f(\omega_2) < -(g(\omega_1) - g(\omega_2)),$$

which implies (taking into account the assertion (ii)) that

$$f(\omega_1) - f(\omega_2) \leq 0 \text{ and } g(\omega_1) - g(\omega_2) \leq 0,$$

equivalently,

$$u(z_1) \leq u(z_2) \text{ and } v(z_1) \leq v(z_2).$$

Passing to the continuity of u and v on $(f + g)(\Omega)$, we start noticing the inequality

$$u(z) \leq u(z + h) \leq u(z) + h \quad \text{for all } z, z + h \in (f + g)(\Omega) \text{ with } h > 0; \tag{2.17}$$

here the first inequality is motivated by the monotony of u, the second follows from the following string of inequalities,

$$z + h = u(z + h) + v(z + h) \geq u(z + h) + v(z) = u(z + h) + z - u(z).$$

In a similar way,

$$u(z) - h \leq u(z - h) \leq u(z) \quad \text{for all } z, z - h \in (f + g)(\Omega) \text{ with } h > 0. \tag{2.18}$$

Combining (2.17) and (2.18), we obtain the continuity of u on $(f + g)(\Omega)$. Replacing u by v, we infer the continuity of v on $(f + g)(\Omega)$.

The next step is to extend u and v by continuity on $\overline{(f + g)(\Omega)}$. This makes use of the monotonicity of these functions.

The proof ends by extending these functions to $\mathbb{R}$, by linearity across each connected component of $\mathbb{R}\backslash\overline{(f + g)(\Omega)}$, so to preserve the relation $u(z) + v(z) = z$. ■

Theorem 2.5.4 (Comonotonic Additivity of Choquet Integral) *Suppose that the functions $f, g \in \mathcal{M}_b(\Omega, \mathcal{A})$ are comonotonic and $f + g \in \mathcal{M}_b(\Omega, \mathcal{A})$. Then*

$$(C) \int_\Omega (f + g)\, \mathrm{d}\mu = (C) \int_\Omega f\, \mathrm{d}\mu + (C) \int_\Omega g\, \mathrm{d}\mu.$$

The argument of Denneberg for Theorem 2.5.4 combines Lemma 2.5.3 with some considerations concerning the pseudo-inverse $\check{S}_f$ of the survival function S_f. It is shown that

$$\check{S}_f + \check{S}_g = \check{S}_{f+g}$$

when f and g are comonotonic, a fact that implies the assertion of Theorem 2.5.4. See [51, Corollary 4.6 p.57 and Proposition 5.1(vi), p. 64].

Marinacci and Montrucchio [124] provided a different argument, based on the theory of finitely additive set functions. See Theorem 2.5.9 below.

Our proof of Theorem 2.5.4 (indicated below) is based on the argument of Dellacherie [49], which reduces the proof to the case of simple functions. We shall need the following approximation result, which we present here in full generality.

Lemma 2.5.5 (Approximation by Simple Functions) *For every $\mathcal{A}$-measurable function $f : \Omega \to \mathbb{R}_+ \cup \{\infty\}$, there exists an increasing sequence $(f_n)_n$ of nonnegative simple functions such that $f(x) = \lim_{n\to\infty} f_n(x)$ for every $x \in \Omega$.*

Moreover, if $f \in \mathcal{M}_b^+(\Omega, \mathcal{A})$ and $g \in \mathcal{M}_b^+(\Omega, \mathcal{A})$ is a function comonotonic to f, then one can choose the approximant sequence $(f_n)_n$ to be uniformly convergent to f and constituted of simple functions comonotonic to g, such that $\{x : f_n(x) > 0\} = \{x : f(x) > 0\}$ for every n.

Proof Define $f_n(x)$ by rounding $f(x)$ down to the nearest integer multiple of 2^{-k} as long $f(x) < n$ and letting $f_n(x) = n$ if $f(x) \geq n$, i.e.,

$$f_n(x) = \begin{cases} \frac{k}{2^n} \text{ if } \frac{k}{2^n} \leq f(x) < \frac{k+1}{2^n},\ k = 0, 1, \ldots, n2^n - 1 \\ n \text{ if } f(x) \geq n. \end{cases}$$

Clearly, the functions f_n are simple, nonnegative and bounded above by f.
Denote $A_{n,k} = \left\{x : \frac{k}{2^n} \leq f(x) < \frac{k+1}{2^n}\right\}$. Then

$$\begin{aligned} A_{n,k} &= \left\{x : \frac{2k}{2^{n+1}} \leq f(x) < \frac{2k+2}{2^{n+1}}\right\} \\ &= A_{n+1,2k} \cup A_{n+1,2k+1}. \end{aligned}$$

For $x \in A_{n+1,2k}$, we have $f_n(x) = \frac{k}{2^n} = \frac{2k}{2^{n+1}} = f_{n+1}(x)$, while for $x \in A_{n+1,2k+1}$ we have $f_n(x) = \frac{k}{2^n} < \frac{2k+1}{2^{n+1}} = f_{n+1}(x)$. Therefore, $(f_n)_n$ is an increasing sequence.

A moment's reflection will show that

$$0 \leq f(x) \leq n \text{ implies } |f(x) - f_n(x)| \leq 2^{-n}.$$

Thus, $\lim_{n\to\infty} f_n(x) = f(x)$ for every x if f is a finite function.

If $f(x) = \infty$, then $f_n(x) = n$ for every n, whence $f_n(x) \to f(x)$.

If f is bounded, say $0 \leq f(x) \leq M$ for every x, then

$$\sup_{x\in\Omega} |f(x) - f_n(x)| \leq 2^{-n} \quad \text{for } n \geq M$$

and thus, the convergence of $(f_n)_n$ to f is uniform. Clearly,

$$\{x : f_n(x) > 0\} = \{x : f(x) > 0\} \quad \text{for every } n.$$

The assertion concerning the property of comonotonicity follows from Lemma 2.5.3 (iii). ■

One can extend Lemma 2.5.5 to the case of extended real-valued measurable functions f by writing $f = f^+ - f^-$, where f^+ and f^- are the positive and negative parts of f. The complex-valued functions can be treated similarly by breaking into real and imaginary parts.

Let us note in passing the following consequence of Lemma 2.5.5 that outlines an interesting connection between the Lebesgue measurable functions and the continuous functions.

Corollary 2.5.6 (Luzin's Theorem) *Let $f : [a,b] \to \mathbb{C}$ be a Lebesgue measurable function. Given any $\varepsilon > 0$, there exists a continuous function $g : [a,b] \to \mathbb{C}$ such that $m(\{x : f(x) \neq g(x)\}) < \varepsilon$.*

The Lipschitz continuity of the Choquet integral (see Proposition 2.4.5) together with Lemma 2.5.5 and Example 2.4.9 provide a useful technique of computing the Choquet integrals by approximating them with integrals of simple functions.

Remark 2.5.7 Every nonnegative simple function $f \in \mathcal{M}_b^+(\Omega, \mathcal{A})$ can be represented as $f = \sum_{k=1}^{n} \alpha_k \chi_{A_k}$, where $\alpha_1, \ldots, \alpha_n \geq 0$ and $A_1 \supset \cdots \supset A_n$ are sets in $\mathcal{A}$. In this case the sets $\{x : f(x) \geq t\}$ with $t \in \mathbb{R}_+$ are precisely $\emptyset, A_n, A_{n-1}, \ldots, A_1$ and Ω.

Proof of Theorem 2.5.4 According to the property of translation invariance, we may reduce ourselves to the case when both functions f and g are nonnegative. Taking into account (in this order) Lemma 2.5.5, Remark 2.5.7 and Lemma 2.5.3 (iii), we may also assume that these functions are simple and admit representations of the form

$$f = \sum\nolimits_{k=1}^{n} \alpha_k \chi_{A_k} \quad \text{with } 0 \leq \alpha_1 \leq \cdots \leq \alpha_n$$

and

$$g = \sum\nolimits_{k=1}^{n} \beta_k \chi_{A_k} \quad \text{with } 0 \leq \beta_1 \leq \cdots \leq \beta_n.$$

Or, in this case,

$$\begin{aligned}(C)\int_\Omega f\,\mathrm{d}\mu + (C)\int_\Omega g\,\mathrm{d}\mu &= \alpha_1\mu(A_1) + \sum\nolimits_{i=1}^{n-1}(\alpha_{i+1} - \alpha_i)\mu(A_i)\\ &\quad + \beta_1\mu(A_1) + \sum\nolimits_{i=1}^{n-1}(\beta_{i+1} - \beta_i)\mu(A_i),\end{aligned}$$

while

$$(C)\int_\Omega (f+g)\,\mathrm{d}\mu = (\alpha_1 + \beta_1)\,\mu(A_1) + \sum\nolimits_{i=1}^{n-1}(\alpha_{i+1} + \beta_{i+1} - \alpha_i - \beta_i)\mu(A_i).$$

The proof is done. ∎

Remark 2.5.8 As noticed Bassanezi and Greco [19], the property of comonotonic additivity is the best we can expect from the additivity of Choquet functionals. Indeed, if the algebra $\mathcal{A}$ contains all singletons and f and g are two functions in $\mathcal{M}_b(\Omega, \mathcal{A})$ such that $f + g \in \mathcal{M}_b(\Omega, \mathcal{A})$ and

$$(C)\int_\Omega (f+g)\,\mathrm{d}\mu = (C)\int_\Omega f d\mu + (C)\int_\Omega g\,\mathrm{d}\mu$$

for every strongly superadditive capacity $\mu : \mathcal{A} \to \mathbb{R}_+$, then necessarily f and g are comonotonic.

Indeed, if f and g were not comonotonic, then would exist points x and y in Ω such that $(f(x) - f(y))\,(g(x) - g(y)) < 0$. Suppose, to make a choice that $f(x) < f(y)$ and $g(x) > g(y)$, and consider the capacity

$$c_{\{x,y\}}(A) = \begin{cases} 1 \text{ if } \{x, y\} \subset A \\ 0 \text{ otherwise.} \end{cases}$$

According to Example 2.4.8,

$$\begin{aligned}(C)\int_\Omega (f+g)\,\mathrm{d}c_{\{x,y\}} &= \inf\{(f+g)(x), (f+g)(y)\} > f(x) + g(y),\\ &= (C)\int_\Omega f\,\mathrm{d}c_{\{x,y\}} + (C)\int_\Omega g\,\mathrm{d}c_{\{x,y\}},\end{aligned}$$

which contradicts the hypothesis.

An interesting consequence of the comonotonic additivity of the Choquet integral is that when integrating comonotonic functions with respect to some capacity μ, their Choquet integrals turn out to be equal to standard integrals with respect to the same probability measure. The following theorem combines results due to Denneberg [51, Proposition 2.10, p. 30], and Ghirardato [84] (Lemma 1, p. 267). The proof given here follows the argument of Lemma 28 in the paper [124] of Marinacci and Montrucchio.

Theorem 2.5.9 *Suppose that $(\Omega, \mathcal{A}, \mu)$ is a capacity space and $\mathcal{F}$ is a family of $\mathcal{A}$-measurable, comonotonic and bounded functions from Ω into $\mathbb{R}$. Then one can find a finitely additive measure ν on $\mathcal{A}$ such that*

$$(C)\int_\Omega f\,\mathrm{d}\mu = \int_\Omega f\,\mathrm{d}\nu,$$

for every $f \in \mathcal{F}$.

Proof For every function $f \in \mathcal{F}$, let $\mathcal{U}_f$ be the family of all upper sets in $\mathcal{A}$ of the form $\{x : f(x) \geq t\}$ or $\{x : f(x) > t\}$ for some $t \in \mathbb{R}$. Clearly, $\mathcal{U}_f$ is a chain. According to Lemma 2.5.3, the union of these families, $\mathcal{U} = \bigcup_{f\in\mathcal{F}} \mathcal{U}_f$, is also a chain.

The next step is to consider a finite chain $\emptyset = A_0 \subset A_1 \subset \cdots \subset A_n \subset A_{n+1} = \Omega$. This chain generates a finite subalgebra $\mathcal{A}_0$ of $\mathcal{A}$, whose atoms are the sets $A_{i+1}\backslash A_i$ with $i \in \{1, \ldots, n\}$. Then the formula

$$\nu_0 \left(A_{i+1} \backslash A_i \right) = \mu \left(A_{i+1} \right) - \mu \left(A_i \right), \quad i \in \{1, \ldots, \},$$

generates a finitely additive measure $\nu_0 : \mathcal{A}_0 \to \mathbb{R}_+$. By a standard extension theorem for finitely additive measures (see Rao and Rao [156], Corollary 3.3.4), there exists a finitely additive measure $\nu \in ba^+ (\mathcal{A})$ that extends ν_0.

Now, let $\mathcal{C}$ be any chain. Let $(\mathcal{C}_\alpha)_\alpha$ be the collection of all its finite subchains, and consider the sets

$$\Gamma_\alpha = \{\lambda \in ba^+ (\mathcal{A}) : \lambda(A) = \mu(A) \text{ for every } A \in \mathcal{C}_\alpha\}.$$

By what we just proved, each Γ_α is nonempty. Moreover, the collection $(\Gamma_\alpha)_\alpha$ has the finite intersection property. Indeed, let $(\mathcal{C}_i)_{i=1}^n \subset (\mathcal{C}_\alpha)_\alpha$ be a finite collection. Since $\bigcup_{i=1}^n \mathcal{C}_i$ is in turn a finite chain, by proceeding as before one can establish the existence of a $\lambda \in ba^+ (\mathcal{A})$ such that $\lambda(A) = \mu(A)$ for every $A \in \bigcup_{i=1}^n \mathcal{C}_i$. As $\lambda \in \bigcap_{i=1}^n \Gamma_i$, the intersection $\bigcap_{i=1}^n \Gamma_i$ is nonempty as desired.

Each Γ_a is a weak*-closed subset of the weak*-compact set

$$\left\{\lambda : \lambda \in ba^+(\mathcal{A}),\ \lambda(\Omega) = \mu(\Omega)\right\}$$

Since $(\Gamma_\alpha)_\alpha$ has the finite intersection property, we conclude that $\bigcap_\alpha \Gamma_\alpha \neq \emptyset$. Any finitely additive set function $\nu \in \bigcap_\alpha \Gamma_\alpha$ verifies the conclusion of the theorem. This ends the proof of the theorem. ■

Based on Theorem 2.5.9, several integral inequalities valid for the Lebesgue integral can be extended to the framework of Choquet integral. However, the requirement concerning comonotonicity can impose some limitations.

So is the case of Chebyshev's algebraic inequality. The original version of this inequality was published by him in [40] and concerned the case of Riemann integrable functions, which are both increasing or decreasing. One year later, Andréief [15] showed that the proper hypothesis was what we call now the comonotonic monotonicity.

Theorem 2.5.10 (Chebyshev's Algebraic Inequality) *Suppose that $\mathcal{A}$ is an algebra of subsets of a set Ω and $f, g \in \mathcal{M}_b^+(\Omega, \mathcal{A})$. Then the following assertions are equivalent:*

(i) *The functions f and g are comonotonic;*
(ii) *The Chebyshev inequality,*

$$\left((C) \int_\Omega f \, d\mu\right) \left((C) \int_\Omega g \, d\mu\right) \leq (C) \int_\Omega fg \, d\mu,$$

holds for all normalized capacities $\mu : \mathcal{A} \to \mathbb{R}_+$.

In the context of probability spaces, this equivalence works for all comonotonic functions $f, g \in \mathcal{M}_b(\Omega, \mathcal{A})$. See [137].

Proof The implication (i) $\Rightarrow$ (ii) follows from Theorem 2.5.9 (and the classical Chebyshev inequality), due to the fact that the functions $1, f, g$ and fg are pairwise comonotonic. Here, the positivity of the functions f and g is essential.

Notice that the implication (i) $\Rightarrow$ (ii) can be easily established in the footsteps of the original inequality. Indeed, according to our hypothesis, for all $x, y \in \Omega$,

$$(f(x) - f(y))(g(x) - g(y)) = f(x)g(x) - f(x)g(y) - f(y)g(x) + f(y)g(y) \geq 0,$$

equivalently,

$$f(x)g(x) + f(y)g(y) \geq f(x)g(y) + f(y)g(x).$$

Integrating this inequality side by side with respect to x and taking into account that the functions $1, f, g$ and fg are pairwise comonotonic, and μ is a a normalized capacity, we obtain

$$(C)\int_\Omega f(x)g(x)\,\mathrm{d}\mu + f(y)g(y) \geq \left((C)\int_\Omega f\,\mathrm{d}\mu\right)g(y) + f(y)\left((C)\int_\Omega g\,\mathrm{d}\mu\right)f(y),$$

for all $y \in \Omega$. The proof ends by integrating both sides of the last inequality with respect to y.

As concerns the converse implication (ii) $\Rightarrow$ (i), notice that the Chebyshev inequality in the case of a discrete measure of the form $\mu = \frac{1}{2}\delta_x + \frac{1}{2}\delta_y$ reads as follows,

$$\begin{aligned}\frac{f(x)+f(y)}{2}\frac{g(x)+g(y)}{2} &= \left((C)\int_\Omega f\,\mathrm{d}\mu\right)\left((C)\int_\Omega g\,\mathrm{d}\mu\right) \leq (C)\int_\Omega fg\,\mathrm{d}\mu \\ &= \frac{f(x)g(x)+f(y)g(y)}{2},\end{aligned}$$

which implies that $(f(x) - f(y))\,(g(x) - g(y)) \geq 0$. Since the points x and y were arbitrarily chosen, we conclude that the functions f and g are comonotonic. ■

It is worth noticing that the restriction on the positivity of the functions f and g cannot be dropped in Theorem 2.5.10. Here is an example due to Girotto and Holzer [87, p. 1121]. They consider the normalized capacity μ defined on the algebra $\mathcal{A} = \mathcal{P}([0, 1])$ by the formula

$$\mu(A) = \inf u\left(\complement A\right) \text{ if } A \neq [0, 1] \text{ and } \mu\left([0, 1]\right) = 1,$$

where

$$u(x) = \begin{cases} 1 & \text{if } x = 0 \\ \frac{3-2x}{4} & \text{if } x \in (0, 1/2) \\ \frac{1-x}{2} & \text{if } x \in [1/2, 1]. \end{cases}$$

Then the functions

$$f(x) = \begin{cases} -2 & \text{if } x \in [0, 1/4) \\ 5 & \text{if } x \in [1/4, 1] \end{cases} \text{ and } g(x) = \begin{cases} -2 & \text{if } x \in [0, 1/2) \\ 1 & \text{if } x \in [1/2, 1] \end{cases}$$

are increasing (and thus comonotonic), but

$$(C)\int_\Omega f\,\mathrm{d}\mu = \frac{19}{8},\ (C)\int_\Omega g\,\mathrm{d}\mu = -\frac{1}{2} \text{ and } (C)\int_\Omega fg\,\mathrm{d}\mu = -\frac{5}{2},$$

which shows that the inequality of Chebyshev does not work for signed functions.

For other extensions of the classical inequalities via Theorem 2.5.9, see Mesiar, Li and Pap [128].

Few words on comonotonicity and the Fubini-type theorems. For functions of two variables, Ghirardato [84] has introduced the following concept of comonotonicity:

Definition 2.5.11 A function $f : \Omega_1 \times \Omega_2 \to \mathbb{R}$ is slice-comonotonic if for every pair of elements $x_1, x_2 \in \Omega_1$, the functions $f(x_1, \cdot)$ and $f(x_2, \cdot)$ are comonotonic and for every pair of elements $y_1, y_2 \in \Omega_2$, the functions $f(\cdot, y_1)$ and $f(\cdot, y_2)$ are comonotonic.

For example, the functions that are monotone in each argument are slice-comonotonic.

The next lemma, also due to Ghirardato, outlines why slice-comonotonicity can be considered as a necessary condition to obtain Fubini-type theorems for non-additive measures.

Lemma 2.5.12 *Let (Ω_1, Σ_1) and (Ω_2, Σ_2) be two measure spaces, and let $f : \Omega_1 \times \Omega_2 \to \mathbb{R}$ be a bounded Σ-measurable mapping, where $\Omega_1 \times \Omega_2$ is endowed with the product σ-algebra $\Sigma = \Sigma_1 \otimes \Sigma_2$. Then, in order that the iterated integrals $\int_{\Omega_2}\left(\int_{\Omega_1} f\,\mathrm{d}\nu_1\right)\mathrm{d}\nu_2$ and $\int_{\Omega_1}\left(\int_{\Omega_2} f\,\mathrm{d}\nu_2\right)\mathrm{d}\nu_1$ exist and are equal for any pair of capacities $\nu_1 : \Sigma_1 \to \mathbb{R}_+$, and $\nu_2 : \Sigma_2 \to \mathbb{R}_+$, it is necessary that f be slice-comonotonic.*

Based on the concept of slice-comonotonicity, Chateauneuf and Lefort [38] have proved the following Fubini-type theorem for non-additive measures on product σ-algebras.

Theorem 2.5.13 *Under the hypotheses of Lemma 2.5.12, let $\nu_1 : \Sigma_1 \to \mathbb{R}_+$, and $\nu_2 : \Sigma_2 \to \mathbb{R}_+$ a pair of strongly subadditive or strongly superadditive capacities, both lower* *and upper continuous. If* $f : \Omega_1 \times \Omega_2 \to \mathbb{R}$ *is a slice-comonotonic, bounded and* Σ*-measurable mapping, then:*

(i) $f(\cdot, \omega_2)$ *is* Σ_1*-measurable and* $\omega_2 \in \Omega_2 \to (C) \int_{\Omega_1} f(\cdot, \omega_2) d\nu_1$ *is bounded and* Σ_2*-measurable.*
(ii) $f(\omega_1, \cdot)$ *is* Σ_2*-measurable and* $\omega_1 \in \Omega_1 \to (C) \int_{\Omega_2} f(\omega_1, \cdot) d\nu_2$ *is bounded and* Σ_1*-measurable.*
(iii) *The iterated integrals* $\iint f d\nu_1 d\nu_2$ *and* $\iint f d\nu_2 d\nu_1$ *exist and are equal i.e.,*

$$\int_{\Omega_2} \left(\int_{\Omega_1} f(\omega_1, \omega_2) d\nu_1 \right) d\nu_2 = \int_{\Omega_1} \left(\int_{\Omega_2} f(\omega_1, \omega_2) d\nu_2 \right) d\nu_1 .$$

Chateauneuf and Lefort [38] have also proved a "full" Fubini-type theorem showing that $\int f \, \mathrm{d}\nu = \iint f \mathrm{d}\nu_1 \mathrm{d}\nu_2 = \iint f \, \mathrm{d}\nu_2 \, \mathrm{d}\nu_1$ for every product capacity $\nu : \Sigma \to \mathbb{R}_+$, with marginals ν_1 and ν_2, which has the Fubini property (in the sense of Ghirardato [84, Definition 5, p. 270]).

2.6 The Subadditivity of Choquet Integral

Subadditivity plays an important role in many areas of mathematics and that motivates the interest in finding the optimal conditions under which the Choquet integral is a subadditive functional.

Theorem 2.6.1 (The Subadditivity Theorem) *In the context of capacities, the following assertions are equivalent:*

(i) $\mu : \mathcal{A} \to \mathbb{R}_+$ *is a strongly subadditive capacity i.e.,*

$$\mu(A \cup B) + \mu(A \cap B) \leq \mu(A) + \mu(B)$$

for all $A, B \in \mathcal{A}$;
(ii) *the Choquet integral associated with* μ *is a subadditive functional i.e.,*

$$(C) \int_{\Omega} (f + g) \, \mathrm{d}\mu \leq (C) \int_{\Omega} f \, \mathrm{d}\mu + (C) \int_{\Omega} g \, \mathrm{d}\mu$$

for all functions $f, g \in \mathcal{M}_b(\Omega, \mathcal{A})$ *such that* $f + g \in \mathcal{M}_b(\Omega, \mathcal{A})$;
(iii) *the Choquet integral associated with* μ *is a submodular functional in the sense that*

$$(C) \int_{\Omega} f \vee g \, \mathrm{d}\mu + (C) \int_{\Omega} f \wedge g \, \mathrm{d}\mu \leq (C) \int_{\Omega} f \, \mathrm{d}\mu + (C) \int_{\Omega} g \, \mathrm{d}\mu$$

for all functions $f, g \in \mathcal{M}_b(\Omega, \mathcal{A})$;

(iv) *the Choquet integral associated with* μ *is a convex functional in the sense that*

$$(\mathrm{C})\int_\Omega ((1-\alpha)f+\alpha g)\,\mathrm{d}\mu \le (1-\alpha)\cdot(\mathrm{C})\int_\Omega f\,\mathrm{d}\mu + \alpha\cdot(\mathrm{C})\int_\Omega g\,\mathrm{d}\mu$$

for all $\alpha \in [0,1]$ *and all functions* $f, g \in \mathcal{M}_b(\Omega, \mathcal{A})$ *such that* $(1-\alpha)f+\alpha g$ *belongs to* $\mathcal{M}_b(\Omega, \mathcal{A})$.

Proof (i) $\Rightarrow$ (ii) The translation invariance of the Choquet integral allows us to reduce ourselves to the case when $f, g \in \mathcal{M}_b^+(\Omega, \mathcal{A})$. Even more, taking into account the approximation Lemma 2.5.5 and the Lipschitz continuity of the integrals (see Proposition 2.4.5), we may assume in addition that g is a simple function. If $g = \alpha\chi_A$ for some $\alpha \ge 0$ and $A \in \mathcal{A}$, then

$$\{x : (f+\alpha\chi_A)(x) \ge t\} = \{x : f(x) \ge t\} \cup (\{x : f(x) \ge t-\alpha\} \cap A)$$

and the strong subadditivity of μ implies that

$$\begin{aligned}\mu(\{x : (f+\alpha\chi_A)(x) \ge t\}) &\le \mu(\{x : f(x) \ge t\}) + \mu(\{x : f(x) \ge t-\alpha\} \cap A)\\ &\quad - \mu(\{x : f(x) \ge t\} \cap (\{x : f(x) \ge t-\alpha\} \cap A)),\end{aligned}$$

whence by integration over $[0, \infty)$, we obtain

$$\begin{aligned}(\mathrm{C})\int_\Omega (f+\alpha\chi_A)\,\mathrm{d}\mu &\le (\mathrm{C})\int_\Omega f\,\mathrm{d}\mu + \int_0^\infty \mu(\{x : f(x) \ge t-\alpha\} \cap A)\,\mathrm{d}t\\ &\quad - \int_0^\infty \mu(\{x : f(x) \ge t\} \cap A)\,\mathrm{d}t\\ &= (\mathrm{C})\int_\Omega f\,\mathrm{d}\mu + \int_{-\alpha}^0 \mu(\{x : f(x) \ge t\} \cap A)\,\mathrm{d}t\\ &= (\mathrm{C})\int_\Omega f\,\mathrm{d}\mu + \alpha\mu(A).\end{aligned}$$

In the general case of simple functions, g is of the form $g = \sum_{i=1}^n \alpha_i\chi_{A_i}$, and we have to proceed by recurrence, starting from

$$f + g = \left(f + \sum_{i=2}^n \alpha_i\chi_{A_i}\right) + \alpha_1\chi_{A_1}.$$

(ii) $\Rightarrow$ (i) Given two sets $A, B \in \mathcal{A}$, the functions $\chi_{A\cup B}$ and $\chi_{A\cap B}$ are comonotonic, so that

$$
\begin{aligned}
\mu(A \cup B) + \mu(A \cap B) &= (\mathrm{C}) \int_{\Omega} \chi_{A \cup B} \, \mathrm{d}\mu + (\mathrm{C}) \int_{\Omega} \chi_{A \cap B} \, \mathrm{d}\mu \\
&= (\mathrm{C}) \int_{\Omega} (\chi_{A \cup B} + \chi_{A \cap B}) \, \mathrm{d}\mu = (\mathrm{C}) \int_{\Omega} (\chi_A + \chi_B) \, \mathrm{d}\mu \\
&\leq (\mathrm{C}) \int_{\Omega} \chi_A \, \mathrm{d}\mu + (\mathrm{C}) \int_{\Omega} \chi_B \, \mathrm{d}\mu \\
&= \mu(A) + \mu(B).
\end{aligned}
$$

(i) $\Rightarrow$ (iii) Since the Choquet integral is translation invariant, it suffices to consider the case when both functions f and g are nonnegative. Then

$$
\begin{aligned}
\{x : (f \vee g)(x) \geq t\} &= \{x : f(x) > t\} \cup \{x : g(x) \geq t\} \\
\{x : (f \wedge g)(x) \geq t\} &= \{x : f(x) \geq t\} \cap \{x : g(x) \geq t\}
\end{aligned}
$$

for every $t \geq 0$. Taking into account that μ is strongly subadditive, this implies that

$$
\begin{aligned}
&\mu(\{x : (f \vee g)(x) \geq t\}) + \mu(\{x : (f \wedge g)(x) \geq t\}) \\
&\qquad \leq \mu(\{x : f(x) \geq t\}) + \mu(\{x : g(x) \geq t\}).
\end{aligned}
$$

As a consequence,

$$
\begin{aligned}
&(\mathrm{C}) \int_{\Omega} f \vee g \, \mathrm{d}\mu + (\mathrm{C}) \int_{\Omega} f \wedge g \, \mathrm{d}\mu \\
&= \int_0^{\infty} \mu(\{x : (f \vee g)(x) \geq t\}) \, \mathrm{d}t + \int_0^{\infty} \mu(\{x : (f \wedge g)(x) \geq t\}) \, \mathrm{d}t \\
&= \int_0^{\infty} [\mu(\{x : (f \vee g)(x) \geq t\}) + \mu(\{x : (f \wedge g)(x) \geq t\})] \, \mathrm{d}t \\
&\leq \int_0^{\infty} [\mu(\{x : f(x) \geq t\}) + \mu(\{x : g(x) \geq t\})] \, \mathrm{d}t \\
&= (\mathrm{C}) \int_{\Omega} f \, \mathrm{d}\mu + (\mathrm{C}) \int_{\Omega} g \, \mathrm{d}\mu,
\end{aligned}
$$

which ends the proof of the implication (i) $\Rightarrow$ (iii).

Notice that the implication (i) $\Rightarrow$ (iii) can be deduced from (i) $\Rightarrow$ (ii) and the fact that (ii) $\Rightarrow$ (iii), due to the formula

$$
f + g = f \vee g + f \wedge g.
$$

For (iii) $\Rightarrow$ (i), notice that for $f = \chi_A$ and $g = \chi_B$, we have $f \vee g = \chi_{A \cup B}$ and $f \wedge g = \chi_{A \cap B}$, so in this case the assertion (iii) coincides with the condition of strong subadditivity of μ.

Finally, the equivalence of the assertions (ii) and (iv) follows from the positive homogeneity of the Choquet integral. ■

Since the Choquet integral is a positive homogeneous functional, the result above can be viewed as a characterization of the conditions under which this functional is sublinear. Recall that the domain of this functional is not in general a vector space (except the case when $\mathcal{A}$ is a σ-algebra).

Remark 2.6.2 A similar argument works in the case of strongly superadditive capacities showing its equivalence with the properties of superadditivity, supermodularity, and concavity of the associated Choquet functional. Here, the supermodularity property means that

$$(\mathrm{C})\int_{\Omega} f\vee g\,\mathrm{d}\mu+(\mathrm{C})\int_{\Omega} f\wedge g\,\mathrm{d}\mu\geq(\mathrm{C})\int_{\Omega} f\,\mathrm{d}\mu+(\mathrm{C})\int_{\Omega} g\,\mathrm{d}\mu$$

for all functions $f, g\in\mathcal{M}_b(\Omega,\mathcal{A})$.

Noticing that every additive capacity is both strongly subadditive and strongly superadditive, we obtain that a capacity is additive if and only if this functional is additive in the sense that

$$(\mathrm{C})\int_{\Omega}(f+g)\,\mathrm{d}\mu=(\mathrm{C})\int_{\Omega} f\,\mathrm{d}\mu+(\mathrm{C})\int_{\Omega} g\,\mathrm{d}\mu$$

for all functions $f, g\in\mathcal{M}_b(\Omega,\mathcal{A})$ such that $f+g\in\mathcal{M}_b(\Omega,\mathcal{A})$.

Remark 2.6.3 If μ is a finitely additive signed measure on a σ-algebra $\mathcal{A}$, then its Choquet functional is a linear functional on the Banach lattice $\mathcal{M}_b(\Omega,\mathcal{A})$, and conversely, if the Choquet functional associated with μ is linear, then μ is a finitely additive measure. By a theorem of Hildebrandt [99] and Fichtenholtz and Kantorovich [65], the dual space of $\mathcal{M}_b(\Omega,\mathcal{A})$ is $ba(\Omega,\mathcal{A})$.

The following result clarifies the concept of Choquet integrability for extended real-valued functions.

Theorem 2.6.4 (Integrability of Extended Real-Valued Functions) *Suppose that $\mu:\mathcal{A}\to\mathbb{R}_+$ is a strongly subadditive capacity and $f:\Omega\to\overline{\mathbb{R}}$ is a measurable function. Then the following assertions are equivalent for A an arbitrary measurable set:*

(i) $(C)\int_A |f|\,\mathrm{d}\mu<\infty$;
(ii) $(C)\int_A f^+\,\mathrm{d}\mu<\infty$ *and* $(C)\int_A f^-\,\mathrm{d}\mu<\infty$;
(iii) $(C)\int_A f\,\mathrm{d}\mu\in\mathbb{R}$ *and* $(C)\int_A(-f)\,\mathrm{d}\mu\in\mathbb{R}$.

Proof The implication (i) $\Rightarrow$ (ii) is a consequence of the monotonicity of Choquet integral and the fact that $f^+, f^- \leq |f|$. The converse implication (ii) $\Rightarrow$ (i) follows from the Subadditivity Theorem 2.6.1:

$$(\mathrm{C})\int_A |f|\, \mathrm{d}\mu = (\mathrm{C})\int_A \left(f^+ + f^-\right) \mathrm{d}\mu \leq (\mathrm{C})\int_A f^+\, \mathrm{d}\mu + (\mathrm{C})\int_A f^-\, \mathrm{d}\mu < \infty.$$

(ii) $\Rightarrow$ (iii) Since μ is finite and subadditive, we have $\overline{\mu} \leq \mu$, where $\overline{\mu}$ is the dual of μ. Then $(\mathrm{C})\int_A f^-\, \mathrm{d}\overline{\mu} < \infty$ and according to the formula (2.11),

$$(\mathrm{C})\int_A f\, \mathrm{d}\mu = (\mathrm{C})\int_A f^+\, \mathrm{d}\mu - (\mathrm{C})\int_A f^-\, \mathrm{d}\overline{\mu} \in \mathbb{R}.$$

The fact that $(\mathrm{C})\int_A (-f)\, \mathrm{d}\mu \in \mathbb{R}$ can be proved similarly.

(iii) $\Rightarrow$ (ii) Indeed, the last formula shows that $(\mathrm{C})\int_A f\, \mathrm{d}\mu \in \mathbb{R}$ implies $(\mathrm{C})\int_A f^+\, \mathrm{d}\mu < \infty$ and $(\mathrm{C})\int_A (-f)\, \mathrm{d}\mu \in \mathbb{R}$ implies

$$(\mathrm{C})\int_A f^-\, \mathrm{d}\mu = (\mathrm{C})\int_A (-f)^+\, \mathrm{d}\mu < \infty,$$

which ends the proof. ■

What can be said about a capacity that is just subadditive (and not strongly subadditive)? Here is a nice property, motivated by the property of translation invariance.

Theorem 2.6.5 *Suppose that $(\Omega, \mathcal{A}, \mu)$ is a capacity space, and μ is a subadditive capacity. Then for every measurable function $f \in \mathcal{M}(\Omega, \mathcal{A})$, bounded from below, and every $A, B \in \mathcal{A}$, we have the inequality*

$$(\mathrm{C})\int_{A\cup B} f\, \mathrm{d}\mu \leq (\mathrm{C})\int_A f\, \mathrm{d}\mu + (\mathrm{C})\int_B f\, \mathrm{d}\mu.$$

Proof Indeed, using the property of invariance to translation, we may assume that f is nonnegative, in which case

$$\begin{aligned}(\mathrm{C})\int_{A\cup B} f\, \mathrm{d}\mu &= \int_0^\infty \mu\left(\{x \in A\cup B : |f(x)| \geq t\}\right) \mathrm{d}t \\ &\leq \int_0^\infty \left[\mu\left(\{x \in A : |f(x)| \geq t\}\right) + \mu\left(\{x \in B : |f(x)| \geq t\}\right)\right] \mathrm{d}t \\ &= (\mathrm{C})\int_A f\, \mathrm{d}\mu + (\mathrm{C})\int_B f\, \mathrm{d}\mu.\end{aligned}$$

■

2.7 Computation of Choquet's Integrals

The effectiveness of classical integral calculus is based on several tools, such as the use of primitives, integration by parts, and change of variables. Unfortunately, these are not available in the general setting of Choquet integral, and the numerical methods prevail.

An exception is provided by the case of distorted Lebesgue measures on $\mathcal{B}(\mathbb{R}_+)$ and of monotone integrands on $\mathbb{R}_+$, due to the results of Sugeno [173, 174]. Later, this limitation concerning the integrands was overcome by Ridaoui and Grabisch [157], who considered the decreasing rearrangements.

The main tool used by Sugeno in his papers is the Laplace transform. An explanation for this is provided by the following result.

Lemma 2.7.1 (Sugeno [173]) *Let $\mu : \mathcal{B}(\mathbb{R}_+) \to \mathbb{R}_+$ be a capacity such that $\mu([\tau, t])$ is continuously differentiable with respect to τ on $[0, t]$ for every $t > 0$ and $\mu(\{t\}) = 0$ for every $t \geq 0$. Then for every strictly increasing and continuously differentiable function $f : [0, +\infty) \to \mathbb{R}_+$,*

$$(C)\int_{[0,t]} f \,\mathrm{d}\mu = -\int_0^t \frac{\partial \mu}{\partial \tau}([\tau, t]) f(\tau)\,\mathrm{d}\tau.$$

If μ is a distorted measure of the form $\mu = m_\varphi$, where $\varphi : [0, \infty) \to \mathbb{R}_+$ is an increasing and continuously differentiable function such that $\varphi(0) = 0$, then

$$(C)\int_{[0,t]} f \,\mathrm{d}m_\varphi = \int_0^t \varphi'(t-\tau) f(\tau)\,\mathrm{d}\tau = (\varphi' * f)(t). \tag{2.19}$$

Here $*$ denotes the operation of convolution.

Proof Indeed,

$$\begin{aligned}
(C)\int_{[0,t]} f \,\mathrm{d}\mu &= \int_0^\infty \mu\left(\{t : f(t) \geq s\} \cap [0, t]\right) \mathrm{d}s \\
&= \int_0^{f(0)} \mu\left([0, t]\right) \mathrm{d}s + \int_{f(0)}^{f(t)} \mu\left(\left[f^{-1}(s), t\right]\right) \mathrm{d}s \\
&= \mu\left([0, t]\right) f(0) + \int_{f(0)}^{f(t)} \mu\left(\left[f^{-1}(s), t\right]\right) \mathrm{d}s.
\end{aligned}$$

Using the change of variables $s = f(\tau)$, we infer that

$$\int_{f(0)}^{f(t)} \mu\left([f^{-1}(s), t]\right) \mathrm{d}s = \int_0^t \mu([\tau, t])\, f'(\tau)\, \mathrm{d}\tau$$
$$= \mu(\{t\})\, f(t) - \mu([0, t])\, f(0) - \int_0^t \frac{\partial \mu}{\partial \tau}([\tau, t]) f(\tau)\, \mathrm{d}\tau$$
$$= -\int_0^t \frac{\partial \mu}{\partial \tau}([\tau, t]) f(\tau)\, \mathrm{d}\tau.$$

In the case of distorted measures $\mu = m_\varphi$, we have

$$\frac{\partial \mu}{\partial \tau}([\tau, t]) = \frac{\partial}{\partial \tau}\varphi(t - \tau) = -\varphi'(t - \tau),$$

which motivates the second assertion of the lemma. ■

For example, for $f(t) = e^t$ and $\varphi(t) = t^2$, we infer from Lemma 2.7.1 that

$$(\mathrm{C}) \int_{[0,t]} e^t\, \mathrm{d}m_{t^2} = \int_0^t 2(t - \tau)e^\tau\, \mathrm{d}\tau$$
$$= \left\{2(t - \tau)e^\tau\right|_0^t + 2\int_0^t e^\tau\, \mathrm{d}\tau$$
$$= 2\left(e^t - t - 1\right).$$

One recognizes in (2.19) a convolution product, which leads to the idea of using the Laplace transform to express the result in a very simple way.

Recall that the Laplace transform of a locally integrable function $f(t)$ defined on $\mathbb{R}_+$ is the function $F = \mathcal{L}[f]$ defined by the formula

$$F(s) = \int_0^\infty f(t)e^{-st} dt \quad \text{for } s \in \mathbb{C},\ \operatorname{Re} s > s_0,$$

where s_0 is the infimum of real numbers α for which the limit

$$\lim_{\substack{a \to 0 \\ b \to \infty}} \int_a^b f(t)e^{-\alpha t}\, \mathrm{d}t$$

exists and is finite.

The Laplace transform is a linear invertible operator, which is extremely useful for solving differential equations, due to the following properties:

($\mathcal{L}1$) $\mathcal{L}[\frac{df}{dt}] = sF(s) - f(0)$;

($\mathcal{L}2$) $\mathcal{L}\left[\int_0^t f(t)dt\right] = \frac{F(s)}{s}$;
($\mathcal{L}3$) $\mathcal{L}(\int_0^t f(t-\tau)g(\tau)d\tau) = F(s)G(s)$.

Consider the relation

$$g(t) = (\mathrm{C}) \int_{[0,t]} f \,\mathrm{d}m_\varphi \tag{2.20}$$

under the assumptions of Lemma 2.7.1 for f and φ. Denoting by $F(s)$, $G(s)$ and $\Phi(s)$ the Laplace transforms of f, g and φ respectively, we obtain from (2.19) and the formulas ($\mathcal{L}2$) and ($\mathcal{L}3$) that

$$G(s) = s\Phi(s)F(s)$$

whence

$$g(t) = \mathcal{L}^{-1}[s\Phi(s)F(s)].$$

On the other hand, given g in the Eq. (2.20), one can determine f (the derivative dg/dm_φ, of g with respect to the capacity m_φ) by the formula

$$f(t) = \mathcal{L}^{-1}\left[\frac{G(s)}{s\Phi(s)}\right](t).$$

For example, for $g(t) = e^{\alpha t} - 1$ with $\alpha > 0$, and $\varphi(t) = t + \alpha t^2/2$, one obtains

$$\frac{df}{dm_\varphi}(t) = \cosh(\alpha t).$$

Ridaoui and Grabisch [157] provided an extension of the above results (based on decreasing rearrangements) to calculate Choquet integrals of continuous and nonnegative functions, which are not necessarily monotonic. See Sect. 3.4 for details concerning the decreasing rearrangements. Other convenient computational formulas for continuous functions (without imposing restrictions on their monotonicity) can be found in the paper [143] of Ontkovičová and Torra.

2.8 The Representation Theorem

The Choquet integral associates to every capacity space $(\Omega, \mathcal{A}, \mu)$ a functional,

$$I(f) = (\mathrm{C}) \int_\Omega f \,\mathrm{d}\mu, \quad f \in \mathcal{M}_b(\Omega, \mathcal{A}),$$

which we referred to as the *Choquet functional* associated with μ. Though nonlinear, this functional displays several important properties like comonotonic additivity, increasing monotonicity, and positive homogeneity. It is also sublinear if (and only if) the capacity μ is strongly subadditive.

As was noticed by Schmeidler [164], all comonotonic additive and monotone increasing functionals on $\mathcal{M}_b(\Omega, \mathcal{A})$ are of this form.

Theorem 2.8.1 *Suppose that* $I : \mathcal{M}_b(\Omega, \mathcal{A}) \to \mathbb{R}$ *is a comonotonic additive and monotone increasing functional. Then the formula* $\mu(A) = I(\chi_A)$ *defines a capacity* μ *on* $\mathcal{A}$ *such that* $\mu(\Omega) = I(1_\Omega)$ *and*

$$
\begin{aligned}
I(f) &= (C)\int_\Omega f \, \mathrm{d}\mu \\
&= \int_0^{+\infty} \mu(\{x : f(x) \geq t\}) \, \mathrm{d}t + \int_{-\infty}^0 [\mu(\{x : f(x) \geq t\}) - I(1_\Omega)] \, \mathrm{d}t
\end{aligned}
$$

for every $f \in \mathcal{M}_b(\Omega, \mathcal{A})$.

Moreover, if in addition I *is upper continuous, then* μ *is upper continuous (, and it is the unique upper continuous capacity representing* I*).*

When $\Omega = \{1, \ldots, N\}$ and $\mathcal{A} = \mathcal{P}(\Omega)$, the set $\mathcal{M}_b(\Omega, \mathcal{A})$ can be identified with $\mathbb{R}^N$ and from Theorem 2.8.1, one can derive the following result (see De Campos and Bolaños [47]):

Corollary 2.8.2 *Let* $F : \mathbb{R}^N \to \mathbb{R}$ *be a given function. Then there exists a unique normalized capacity* μ *on* $\mathcal{P}(\{1, \ldots N\})$ *such that* $F(x) = (C)\int_{\{1,\ldots N\}} x \, \mathrm{d}\mu$ *for all* $x \in \mathbb{R}^N$ *if and only if* F *fulfills the following properties:*

(i) *comonotonic additivity;*
(ii) *increasing monotonicity;*
(iii) $F(1_{\{1,\ldots N\}}) = 1$ *and* $F(0) = 0$.

Moreover, μ *is defined by* $\mu(A) = F(\chi_A)$.

The proof of Theorem 2.8.1 needs two auxiliary results, which will be later extended in Sect. 4.4.

Lemma 2.8.3 *Every comonotonic additive and monotone increasing functional* I *defined on* $\mathcal{M}_b(\Omega, \mathcal{A})$ *is positively homogeneous.*

Proof Since I is comonotonic additive, we get $I(0) = I(0+0) = 2 \cdot I(0)$, which implies $I(0) = 0$. As a consequence, $I(0 \cdot f) = 0 = 0 \cdot I(f)$. Then the same argument shows that $I(2f) = I(f+f) = 2I(f)$ and by mathematical induction, we infer that $I(pf) = pI(f)$ for all $p \in \mathbb{N}$.

Now, consider the case of positive rational numbers $r = p/q$, where $p, q \in \mathbb{N}$. Then $I(f) = I\left(q \cdot \frac{1}{q} f\right) = qI\left(\frac{1}{q} f\right)$, which implies $I\left(\frac{1}{q} f\right) = \frac{1}{q} \cdot I(f)$. Therefore,

$$I\left(\frac{p}{q} f\right) = pI\left(\frac{1}{q} f\right) = \frac{p}{q} \cdot I(f).$$

Passing to the case of an arbitrary positive number α, let us choose a decreasing sequence $(r_n)_n$ of rationals converging to α. Then $r_n f \geq \alpha f$, which yields $r_n I(f) = I(r_n f) \geq I(\alpha f)$. Passing here to limit it follows $\alpha I(f) \geq I(\alpha f)$. In a similar way, by considering an increasing sequence of positive rational numbers $(s_n)_n$ that converges to α, one obtains $\alpha I(f) \leq I(\alpha f)$, which combined with the previous inequality proves the assertion of Lemma 2.8.3 in the case of nonnegative functions.

When $f \in \mathcal{M}_b(\Omega, \mathcal{A})$ is arbitrary, one can choose a positive number λ such that $f + \lambda \geq 0$. By the above reasoning and the property of comonotonic additivity, for all $\alpha \geq 0$,

$$\begin{aligned}\alpha I(f) + \alpha\lambda I(1_\Omega) &= \alpha I\left(f + \lambda 1_\Omega\right) = I(\alpha\left(f + \lambda 1_\Omega\right)) \\ &= I(\alpha f + \alpha\lambda 1_\Omega) = I(\alpha f) + \alpha\lambda I(1_\Omega),\end{aligned}$$

which ends the proof of Lemma 2.8.3. ∎

Lemma 2.8.4 *Every comonotonic additive and monotone increasing functional I defined on $\mathcal{M}_b(\Omega, \mathcal{A})$ is Lipschitz continuous.*

Proof Given $f, g \in \mathcal{M}_b(\Omega, \mathcal{A})$, one can choose a decreasing sequence $(\alpha_n)_n$ of positive numbers such that $\alpha_n \downarrow \|f - g\|$. Then $f \leq g + \|f - g\| \leq g + \alpha_n \cdot 1_\Omega$, which implies

$$I(f) \leq I(g + \alpha_n \cdot 1_\Omega) = I(g) + \alpha_n I(1_\Omega),$$

due to the properties of monotonicity, translation invariance, and positive homogeneity of I. Translation invariance is a consequence of comonotonic additivity while positive homogeneity follows from Lemma 2.8.3. Since the roles of f and g are symmetric, this leads to the fact that $|I(f) - I(g)| \leq I(1_\Omega) \cdot \alpha_n$ for all n, whence, by passing to the limit as $n \to \infty$, we conclude that

$$|I(f) - I(g)| \leq I(1_\Omega) \cdot \|f - g\|.$$

The proof is complete. ∎

Proof (of Theorem 2.8.1) The fact that μ is a capacity is obvious. According to Lemma 2.8.4 and the property of translation invariance, it suffices to prove the formula stated in Theorem 2.8.1 for the positive simple functions. Every such function f admits a representation of the form

$$f = \sum_{i=1}^{n} \alpha_i \chi_{A_i}$$

with $0 \leq \alpha_n \leq \cdots \leq \alpha_1$ and the sets $A_1, \ldots, A_n \in \mathcal{A}$ are pairwise disjoint and nonempty. According to Example 2.4.9,

$$(\mathrm{C}) \int_{\Omega} f \, \mathrm{d}\mu = \sum_{i=1}^{n} (\alpha_i - \alpha_{i+1}) \mu \left(\bigcup_{j=1}^{i} A_j \right),$$

where $\alpha_{n+1} = 0$ by definition. It remains to prove that also

$$I(f) = \sum_{i=1}^{n} (\alpha_i - \alpha_{i+1}) \mu \left(\bigcup_{j=1}^{i} A_j \right), \tag{2.21}$$

which can be done by induction over n. For $n = 1$, $f = \alpha_1 \chi_{A_1}$ and in this case

$$I(\alpha_1 \chi_{A_1}) = \alpha_1 I(\chi_{A_1}) = \alpha_1 \mu(A_1)$$

since I is positively homogeneous. Assuming the formula (2.21) true for all simple functions f with $k < n$ summands, we will prove that the same happens for n summands. Indeed, we can split f as

$$f = g + h,$$

where

$$g = \sum_{k=1}^{n-1} (\alpha_k - \alpha_n) \chi_{A_k} \text{ and } h = \alpha_n \chi_{\cup_{k=1}^{n} A_k}.$$

According to the induction hypothesis,

$$\begin{aligned} I(g) &= \sum_{i=1}^{n-1} ((\alpha_i - \alpha_n) - (\alpha_{i+1} - \alpha_n)) \mu \left(\bigcup_{j=1}^{i} A_j \right) \\ &= \sum_{i=1}^{n-1} (\alpha_i - \alpha_{i+1}) \mu \left(\bigcup_{j=1}^{i} A_j \right) \end{aligned}$$

and

$$I(h) = \alpha_n \mu \left(\bigcup_{k=1}^{n} A_k \right).$$

Since g and h are comonotonic, it follows that $I(f) = I(g+h) = I(g) + I(h)$ and the proof is done. ■

The possibility to represent a functional as a Choquet integral offers a powerful way to certificate its comonotonic additivity. A surprising example from Real Analysis is presented in what follows.

Example 2.8.5 (Limit Inferior and Limit Superior; cf. Denneberg [51, p.70]) Let $\Omega = \mathbb{N}$ and let H be the strongly subadditive capacity defined on $\mathcal{P}(\mathbb{N})$ via the formula $H(A) = 0$ if A is a finite subset of $\mathbb{N}$ and $H(A) = 1$ otherwise. Then for every bounded sequence $X : \mathbb{N} \to \mathbb{R}$, the following two formulas

$$(\mathrm{C}) \int_{\mathbb{N}} X \, \mathrm{d}H = \limsup_{n \to \infty} X(n) = \inf_{n \geq 1} \sup_{k \geq n} X(k)$$

and

$$(\mathrm{C}) \int_{\mathbb{N}} X \, \mathrm{d}\overline{H} = \liminf_{n \to \infty} X(n) = \sup_{n \geq 1} \inf_{k \geq n} X(k)$$

hold. Here $\overline{H}$ denotes the dual capacity. For the first formula, notice that we may assume (by translation invariance) that $X > 0$ and in this case, take into account that

$$\limsup_{n \to \infty} X(n) < \alpha$$

implies that at most finitely many terms $X(n)$ are bigger than α. See [43, p. 58] for details. The second formula follows from the first one. Consequently,

$$I : \ell^\infty \to \mathbb{R}, \quad I((x_n)_n) = \limsup_{n \to \infty} x_n,$$

constitutes an example of Choquet functional.

Remark 2.8.6 The previous example reveals the striking fact that the limit superior and limit inferior are comonotonic additive. In other words, if

$$(x_m - x_n)(y_m - y_n) \geq 0 \quad \text{for all } m, n \in \mathbb{N},$$

then

$$\limsup_{n\to\infty}(x_n + y_n) = \limsup_{n\to\infty} x_n + \limsup_{n\to\infty} y_n$$

and

$$\liminf_{n\to\infty}(x_n + y_n) = \liminf_{n\to\infty} x_n + \liminf_{n\to\infty} y_n$$

for all bounded sequences $(x_n)_n$ and $(y_n)_n$. A similar result works for the supremum and the infimum of bounded sequences. Use Example 2.4.8.

2.9 A Capacity Analog of the Riesz–Kakutani Representation Theorem

An important result at the frontier between Measure Theory and Functional Analysis is the Riesz–Kakutani representation theorem. According to this theorem, if K is a compact Hausdorff space, then every linear and positive functional I defined on the space $C(K)$ can be represented (uniquely) as

$$I(f) = \int_K f \, \mathrm{d}\mu, \quad f \in C(K),$$

where $\mu : \mathcal{B}(K) \to \mathbb{R}_+$ is a regular Borel measure. Regularity means that

$$\mu(A) = \sup\{\mu(C) : C \text{ closed}, C \subset A\} = \inf\{\mu(O) : O \text{ open}, O \supset A\}$$

for all $A \in \mathcal{B}(K)$,

See Hewitt and Stromberg [98, Theorem 12.36, p. 177] (where this result is presented in the more general context of functions defined on locally compact Hausdorff spaces).

Zhou [195] proved the following companion to the Riesz–Kakutani representation theorem in the setting of Choquet integrals.

Theorem 2.9.1 *Suppose that $I : C(K) \to \mathbb{R}$ is an upper continuous, comonotonic additive and monotone increasing functional such that $I(1) > 0$. Then there exists a unique upper-continuous capacity μ on the lattice $\Sigma_{up}^{+}(K)$ (the collection of all upper contour sets), such that,*

$$I(f) = \int_0^{+\infty} \mu(\{x : f(x) \geq t\}) \, \mathrm{d}t + \int_{-\infty}^{0} [\mu(\{x : f(x) \geq t\}) - I(1))] \, \mathrm{d}t \qquad (2.22)$$

for all $f \in C(K)$,

On the other hand, for any upper-continuous capacity $\mu : \Sigma_{up}^{+}(K) \to \mathbb{R}_{+}$, *the functional* I *defined on* $C(K)$ *by the formula (2.22) with* $I(1) = \mu(K)$, *is an upper continuous, comonotonic additive and monotone increasing functional.*

In Sect. 2.1, the concept of capacity was restricted to the case of set functions defined on algebras of sets. However, it can be extended verbatim to the case of functions defined on lattices (and this applies in the case of Theorem 2.9.1). The concept of Choquet integral can be introduced in the same manner and most of the results presented in Sects. 2.4–2.6 remain true in this enlarged framework. However, we will not enter the details here, mentioning instead, the following particular case of Theorem 2.9.1, due to Epstein and Wang [58]:

Corollary 2.9.2 *Let* K *be a compact metric space and* $I : C(K) \to \mathbb{R}$ *an upper continuous, comonotonic additive and monotone increasing functional such that* $I(1) = 1$. *Then there exists a unique upper continuous normalized capacity* $\mu : \mathcal{B}(K) \to \mathbb{R}_{+}$ *such that*

$$I(f) = (\mathrm{C}) \int_K f \, \mathrm{d}\mu \quad \textit{for all } f \in C(K).$$

Moreover, μ *verifies the following two properties of regularity:*

(IR) (*inner regularity*) $\mu(A) = \sup\{\mu(C) : C \text{ compact}, C \subset A\}$ *for all* A *in* $\mathcal{B}(K)$;
(OR) (*outer regularity*) $\mu(C) = \inf\{\mu(O) : O \text{ open}, O \supset C\}$ *for all closed sets* $C \subset K$.

The derivation of Corollary 2.9.2 from Theorem 2.9.1 is detailed by Zhou [195].

An application of Corollary 2.9.2 to the theory of mass transportation can be found in our paper [74].

2.10 Comments

The Layer Cake Representation extends Cavalieri's principle from the geometric measure theory. See Grigoryan [94, Theorem 2.23, p. 79].

The basics of the non-additive integration on real line appeared for the first time in a paper from 1925 of Vitali [178]. Its importance from a historical perspective is discussed by Marinacci [123].

The abstract notion of capacity (as well as the integral of a nonnegative function with respect to a capacity) was introduced by Choquet [41] in analogy with the physical concept of electrostatic capacity. The whole story is told by Choquet in [42]. See also Landkof [112] and Wermer [185].

Choquet was primarily interested in the capacities $H : \mathcal{P}(\mathbb{R}^N) \to [0, \infty]$, which satisfy the following five conditions:

(Ch1) $H(\emptyset) = 0$;
(Ch2) if $A \subset B$, then $H(A) \leq H(B)$ *(monotonicity)*;
(Ch3) for every sequence of subsets A_n of $\mathbb{R}^N$,

$$H\left(\bigcup\nolimits_n A_n\right) \leq \sum\nolimits_n H(A_n) \quad \textit{(countable subadditivity)};$$

(Ch4) for every non-increasing sequence of compact subsets K_n of $\mathbb{R}^N$,

$$H\left(\bigcap\nolimits_n K_n\right) = \lim_{n\to\infty} H(K_n) \quad \textit{(outer regularity)};$$

(Ch5) for every increasing sequence of subsets A_n of $\mathbb{R}^N$,

$$H\left(\bigcup\nolimits_n A_n\right) = \lim_{n\to\infty} H(A_n) \quad \textit{(inner regularity)}.$$

The main feature of the capacities which satisfy the conditions (Ch1), (Ch2), (Ch4), and (Ch5) is that all Borel subsets E of $\mathbb{R}^N$ are capacitable in the sense that

$$\sup\{H(K) : K \subset E,\ K \text{ compact}\} = \inf\{H(U) : U \supset E,\ U \text{ open}\}.$$

Later on, the concept of capacity was ascribed to all set functions μ defined on an algebra $\mathcal{A}$ of sets, which verify the conditions (Ch1) and (Ch2). See Adams [2], Carleson [31], Denneberg [51] and Wermer [185], to cite just a few from the most prominent promoters of the capacity theory.

The important result in Sect. 2.6 concerning the equivalence between the strong subadditivity of a capacity and the convexity of the Choquet integral associated with it is due to Choquet [41].

The strongly subadditive set functions play an important role in combinatorial optimization. See Lovász [121, 122] and references therein.

The Choquet integral was rediscovered (and extended to the case of signed functions) by Schmeidler [164], who was interested in the applications of this integral to decision-making under risk and uncertainty. See his paper [165] and references therein.

Various integral representations of comonotonic additive functionals were obtained by Greco [93], Schmeidler [164], Cerreia-Vioglio, Maccheroni, Marinacci and Montrucchio [35, 36], and Zhou [195].

The following example exhibits a family of capacities that received a great deal of attention in the last years, providing information about how well a function in a Bessel potential space can be approximated by functions that vanish on a given set.

Remark 2.10.1 (The Bessel Capacities) Let $\alpha > 0$ and $s > 1$ be two parameters. We define the space $H^{\alpha,s}(\mathbb{R}^N)$ $(N \geq 1)$ of *Bessel potentials* as the completion of $C_c^\infty(\mathbb{R}^N)$ with respect to the norm

$$\|u\|_{H^{\alpha,s}} = \left\|(1 - \Delta)^{\alpha/2} u\right\|_{L^s(\mathbb{R}^N)}.$$

Here the operator $(1 - \Delta)^{\alpha/2}$ is understood as $\mathcal{F}^{-1}(1+\|\xi\|^2)^{\alpha/2}\mathcal{F}$, where $\mathcal{F}$ is the Fourier transform in $\mathbb{R}^N$,

$$\mathcal{F}(f)(\xi) = \frac{1}{(2\pi)^{N/2}} \int_{\mathbb{R}^N} f(x)e^{-i\langle x,\xi\rangle}dx.$$

See Grafakos [92] and Vretblad [180] for details on this transform and its applications.

One can prove that a function u belongs to $H^{\alpha,s}(\mathbb{R}^N)$ if and only if

$$u = G_\alpha * f$$

for some $f \in L^s\left(\mathbb{R}^N\right)$, and moreover, $\|u\|_{H^{\alpha,s}(\mathbb{R}^N)} = \|f\|_{L^s(\mathbb{R}^N)}$. See [126]. Here G_α is the *Bessel kernel* of order α defined by

$$G_\alpha(x) = \mathcal{F}^{-1}[(1 + \|\xi\|^2)^{-\alpha/2}](x).$$

The *Bessel capacity* of a nonempty set $E \subset \mathbb{R}^N$ is defined by

$$\operatorname{Cap}_{\alpha,s}(E) = \inf\left\{\|f\|_{L^s}^s : f \geq 0,\ G_\alpha * f \geq 1 \text{ on } E\right\}.$$

By definition $\operatorname{Cap}_{\alpha,s}(\emptyset) = 0$. The Bessel capacity is an outer measure, which verifies in addition the following properties:

- $\operatorname{Cap}_{\alpha,s}(E) = \inf\left\{\operatorname{Cap}_{\alpha,s}(U) : U \text{ open}, U \supset E\right\}$;
- if $(K_n)_n$ is a decreasing sequence of compact sets of $\mathbb{R}^N$, then

$$\operatorname{Cap}_{\alpha,s}\left(\bigcap\nolimits_n K_n\right) = \lim_{n\to\infty} \operatorname{Cap}_{\alpha,s}(K_n);$$

- if $(E_n)_n$ is an increasing sequence of subsets of $\mathbb{R}^N$, then

$$\operatorname{Cap}_{\alpha,s}\left(\bigcup\nolimits_n E_n\right) = \lim_{n\to\infty} \operatorname{Cap}_{\alpha,s}(E_n).$$

For $s = 2$ and $\alpha \in (0, 1]$, (as well as for $\alpha = 1$ and $s > 1$), the Bessel capacity is strongly subadditive. See Ooi and Phuc [146] and references therein.

When dealing with cognitive ambiguity, traditional probability approaches have several shortcomings. Non-probabilistic uncertainty quantification techniques like possibility theory are thus suggested and utilized. See Dubois and Prade [57]. Let us mention that a capacity $\mu : \mathcal{P}(\Omega) \to [0, 1]$ is called a *possibility measure* if it verifies the conditions $\mu(\emptyset) = 0,\ \mu(\Omega) = 1$ and

$$\mu\left(\bigcup\nolimits_{i\in I} A_i\right) = \sup\{\mu(A_i) : i \in I\}$$

for every at most countable family $(A_i)_{i\in I}$ of subsets of Ω. A function λ (from Ω into $[0, 1]$ is called a *possibility distribution* if $\sup\{\lambda(\omega) : \omega \in \Omega\} = 1$. Any possibility distribution λ on Ω induces the possibility measure μ_λ given by the formula $\mu_\lambda(A) = \sup\{\lambda(\omega) : \omega \in A\}$ if $A \neq \emptyset$ and $\mu(\emptyset) = 0$. One can easily show that every possibility measure is strongly subadditive.

A much more general notion than that of finite capacity is the following one, used in Mathematical Economics.

Definition 2.10.2 A game on a set Ω is any set function $\nu : \mathcal{A} \to \mathbb{R}$ defined on an algebra $\mathcal{A}$ of subsets of Ω with the only requirement that $\nu(\emptyset) = 0$.

In terms of games, the capacities are nothing but the monotone games.

From the game-theoretic point of view, Ω can be interpreted as a set of players and $\mathcal{A}$ as the algebra of admissible coalitions in Ω.

The value $\nu(A)$ represents the total payoff that coalition A can obtain (equivalently, the worth of A).

The notions of *positive game, monotone game, dual game, additive game, strongly subadditive game*, etc., can be introduced as in the case of capacities.

Aumann and Shapley [17] have defined the *variation norm* of a game ν by the formula

$$\|\nu\|_{bv} = \sup \sum\nolimits_{i=1}^{n} |\nu(A_i) - \nu(A_{i-1})|$$

where the supremum is taken over all finite chains $\emptyset = A_1 \subset A_2 \subset \cdots \subset A_n = \Omega$. The set of all games with finite variation norm constitutes (with respect of the addition of functions, scalar multiplication and variation norm) a Banach space, denoted $bv(\mathcal{A})$.

Theorem 2.10.3 (Aumann and Shapley) *A game ν belongs to $bv(\mathcal{A})$ if and only if there exist two capacities μ_1 and μ_2 such that $\nu = \mu_1 - \mu_2$ and $\|\nu\|_{bv} = \mu_1(\Omega) + \mu_2(\Omega)$.*

The decomposition of ν is realized by the capacities $\mu_1 = \nu^+$ and $\mu_2 = \nu^-$, defined respectively by the formulas

$$\nu^+(A) = \sup \sum\nolimits_{i=1}^{n} \max\{\nu(A_i) - \nu(A_{i-1}), 0\},$$

where the supremum is taken over all finite chains $\emptyset = A_1 \subset A_2 \subset \cdots \subset A_n = A$, and

$$\nu^-(A) = \nu^+(A) - \nu(A).$$

For full details see [17, Proposition 4.1, p. 27].

It is worth mentioning that $bv(\mathcal{A})$ is actually a Banach lattice when endowed with the positive cone

$$bv^+(\mathcal{A}) = \{\mu : \mu : \mathcal{A} \to \mathbb{R}_+ \text{ is a capacity}\}.$$

Indeed, ν^+ and ν^- represent the positive and negative part of ν and $|\nu| = \nu^+ + \nu^-$ represents the modulus of ν.

The formula defining the Choquet integral of a bounded and measurable function extends verbatim to the case of games with bounded variation. Precisely, given a game $\nu \in bv(\mathcal{A})$ and a function $f \in f \in \mathcal{M}_b(\Omega, \mathcal{A})$, one defines the integral of f over a set $A \in \mathcal{A}$ by the formula

$$\begin{aligned}(\mathrm{C}) \int_A f \,\mathrm{d}\nu = &\int_0^{+\infty} \nu\left(\{x : f(x) \geq t\} \cap A\right) \mathrm{d}t \\ &+ \int_{-\infty}^0 [\nu\left(\{x : f(x) \geq t\} \cap A\right) - \mu(A)] \,\mathrm{d}t.\end{aligned}$$

Both integrals appearing in the right-hand side are Riemann integrals since the integrands

$$\nu(\{x : f(x) \geq t\}) = \nu^+(\{x : f(x) \geq t\}) - \nu^-(\{x : f(x) \geq t\})$$

are functions of bounded variation in t.

Lemma 2.4.2 (about the replacement of sign $\geq$ by $>$) continues to work in this extended framework and the same is true for the uniqueness of the extension of Choquet integral from $\mathcal{M}_b^+(\Omega, \mathcal{A})$ to $\mathcal{M}_b(\Omega, \mathcal{A})$, by preserving the property of translation invariance.

A caution word is necessary for the extension of Proposition 2.4.5, where the properties of positivity and monotonicity hold only for capacities and the property of Lipschitz continuity should be restated in the context of games $\nu \in bv(\mathcal{A})$ as follows:

$$\left|(C) \int_\Omega f \,\mathrm{d}\nu - (C) \int_\Omega g \,\mathrm{d}\nu\right| \leq \|\nu\|_{bv} \|f - g\|$$

for all $f, g \in \mathcal{M}_b(\Omega, \mathcal{A})$.

Notice the availability of the properties of translation invariance, positive homogeneity, and of the formula of asymmetry,

$$(\mathrm{C})\int_A f\,\mathrm{d}\nu = -(\mathrm{C})\int_A (-f)\,\mathrm{d}\bar{\nu},$$

where $\bar{\nu}$ is the game dual to ν.

A straightforward consequence of Theorem 2.5.4, on the comonotonic additivity of Choquet integral associated with a capacity, is the validity of this result also in the framework of games with bounded variation.

The same applies in the case of the Subadditivity Theorem 2.6.1. The extension of the Representation Theorem 2.8.1 in the context of games can be found in a paper by Marinacci and Montrucchio. See [124, Theorem 32].

The Choquet integral is not the only nonadditive integral known in the literature. We mention here two other integrals, which can be attached to every capacity space $(\Omega, \mathcal{A}, \mu)$:

- The *Šipoš integral,* introduced by Šipoš [169, 170] by the formula

$$(S)\int_\Omega f\,\mathrm{d}\mu = (\mathrm{C})\int_\Omega f^+\,\mathrm{d}\mu - (\mathrm{C})\int_\Omega f^-\,\mathrm{d}\mu, \quad f \in \mathcal{M}_b(\Omega, \mathcal{A}).$$

  this integral coincides with the Choquet integrals when $f \geq 0$ and it is not comonotonic additive.
- The *Sugeno integral,* introduced by Sugeno [171, 172] by the formula

$$(Sug)\int_\Omega f\,\mathrm{d}\mu = \bigvee\nolimits_{A\in\mathcal{A}} \left(\bigwedge\nolimits_{x\in A} f(x) \wedge \mu(A)\right), \quad f \in \mathcal{M}_b^+(\Omega, \mathcal{A}).$$

For many other generalizations see Lehrer [113], Mesiar [127], Pap [149] and Zhang, Mesiar and Pap [192–194].

The Lebesgue Capacitary Spaces

3

As noticed in the previous chapter, the Choquet integral need not be linear and in fact may not even be sublinear. The sublinearity of this integral implies the triangle inequality, from which Hölder's and Minkowski's inequalities follow from usual convexity arguments. These inequalities in turn serve as a basis for the study of a family of Banach spaces of functions denoted $L^p(\mu)$, consisting of those measurable functions whose pth power have finite Choquet integrals.

According to Theorem 2.6.1, the property of sublinearity of the integral is equivalent to the strong subadditivity of the capacity under attention. On the other hand, a number of useful tools, like Fatou's lemma, impose the restriction to work with lower continuous capacities. This situation motivates our option to discuss the construction of the analogs of L^p spaces within the framework of strongly subadditive and lower continuous capacities.

3.1 Some Basic Capacitary Inequalities

The Theorem 2.6.1 concerning the subadditivity of the Choquet integral implies a number of important integral inequalities that extends some classical results from real analysis. For technical considerations, in this section, we will restrict ourselves to the case of bounded measurable functions.

Theorem 3.1.1 *If $\mu : \mathcal{A} \to \mathbb{R}_+$ is a strongly subadditive capacity and f and g are two functions such that $f, g, f - g \in \mathcal{M}_b(\Omega, \mathcal{A})$, then for all $A \in \mathcal{A}$,*

$$\left|(\mathrm{C}) \int_A f \,\mathrm{d}\mu - (\mathrm{C}) \int_A g \,\mathrm{d}\mu\right| \le (\mathrm{C}) \int_A |f - g| \,\mathrm{d}\mu.$$

S. G. Gal, C. P. Niculescu, *Choquet Integral and Monotone Sublinear Operators*, Frontiers in Mathematics, https://doi.org/10.1007/978-3-031-71701-7_3

Proof Indeed,

$$(\mathrm{C})\int_A f\,\mathrm{d}\mu = (\mathrm{C})\int_A (f-g+g)\,\mathrm{d}\mu \le (\mathrm{C})\int_A (|f-g|+g)\,\mathrm{d}\mu$$
$$\le (\mathrm{C})\int_A |f-g|\,\mathrm{d}\mu + (\mathrm{C})\int_A g\,\mathrm{d}\mu,$$

whence

$$(\mathrm{C})\int_A f\,\mathrm{d}\mu - (\mathrm{C})\int_A g\,\mathrm{d}\mu \le (\mathrm{C})\int_A |f-g|\,\mathrm{d}\mu.$$

The proof ends by interchanging the roles of f and g. ■

For $g = 0$,, the inequality stated by Theorem 3.1.1 reduces to the capacitary form of the *modulus inequality*, i.e.,

$$\left|(\mathrm{C})\int_A f\,\mathrm{d}\mu\right| \le (\mathrm{C})\int_A |f|\,\mathrm{d}\mu.$$

It is important to mention that the capacitary modulus inequality holds in a much more general context.

Theorem 3.1.2 *If $\mu : \mathcal{A} \to \mathbb{R}_+$ is a strongly subadditive capacity and $f : \Omega \to \overline{\mathbb{R}}$ is an extended real-valued measurable functions such that* (C) $\int_\Omega |f|\,\mathrm{d}\mu < \infty$, *then the Choquet integral of f over any set $A \in \mathcal{A}$ of finite capacity exists and*

$$\left|(\mathrm{C})\int_A f\,\mathrm{d}\mu\right| \le (\mathrm{C})\int_A |f|\,\mathrm{d}\mu.$$

Proof According to Theorem 2.6.4, the Choquet integral of f and $-f$ exist, and they are finite.

If (C) $\int_A f\,\mathrm{d}\mu \ge 0$, then the modulus inequality follows from the monotone increasing property of the integral and the fact that $f \le |f|$.

If (C) $\int_A f\,\mathrm{d}\mu \le 0$,, then

$$(\mathrm{C})\int_A f^{+}\mathrm{d}\mu \le (\mathrm{C})\int_A f^{-}\mathrm{d}\overline{\mu}.$$

See formula (2.11) in Sect. 2.4. According to Theorem 2.6.1, μ is subadditive, which implies that $\overline{\mu} \le \mu$. Hence

$$
\begin{aligned}
0 \leq -(\mathrm{C}) \int_A f \,\mathrm{d}\mu &= (\mathrm{C}) \int_A f^- \,\mathrm{d}\overline{\mu} - (\mathrm{C}) \int_A f^+ \mathrm{d}\mu \\
&\leq (\mathrm{C}) \int_A f^- \,\mathrm{d}\overline{\mu} \leq (\mathrm{C}) \int_A |f| \,\mathrm{d}\mu.
\end{aligned}
$$

In conclusion, $\pm \left((\mathrm{C}) \int_A f \,\mathrm{d}\mu\right) \leq (\mathrm{C}) \int_A |f| \,\mathrm{d}\mu$ and the proof is complete. ∎

Corollary 3.1.3 *Suppose that $\mu : \mathcal{A} \to \mathbb{R}_+$ is a strongly subadditive capacity. Then every function $f \in \mathcal{M}_b(\Omega, \mathcal{A})$ verifies the double inequality*

$$
\inf_{x \in \Omega} f(x) \cdot \mu(\Omega) \leq (\mathrm{C}) \int_\Omega f \,\mathrm{d}\mu \leq \sup_{x \in \Omega} f(x) \cdot \mu(\Omega).
$$

As a consequence,

$$
\left|(\mathrm{C}) \int_\Omega f \,\mathrm{d}\mu\right| \leq \sup_{x \in \Omega} |f(x)| \cdot \mu(\Omega).
$$

A simple, but important inequality, is the following capacitary analogue of Chebyshev's inequality from Probability Theory and Statistics.

Theorem 3.1.4 *If $(\Omega, \mathcal{A}, \mu)$ is a capacity space and $f : \Omega \to \overline{\mathbb{R}}$ is a measurable function, then*

$$
\mu\left(\{x : |f(x)| \geq \alpha\}\right) \leq \frac{1}{\alpha} \cdot (\mathrm{C}) \int_\Omega |f| \,\mathrm{d}\mu
$$

for every $\alpha > 0$.

Proof Indeed,

$$
\begin{aligned}
(\mathrm{C}) \int_\Omega |f| \,\mathrm{d}\mu &= \int_0^\infty \mu\left(\{x : |f(x)| \geq t\}\right) \mathrm{d}t \\
&\geq \int_0^\alpha \mu\left(\{x : |f(x)| \geq \alpha\}\right) \mathrm{d}t \\
&= \alpha\mu\left(\{x : |f(x)| \geq \alpha\}\right).
\end{aligned}
$$

∎

In analogy with the Probability Theory, the measurable functions associated with a normalized capacity space $(\Omega, \mathcal{A}, \mu)$, are also called *random* variables. The *expected value* and the *variance* of a random variable f are defined respectively as

$$\mathrm{E}(f) = (\mathrm{C}) \int_{\Omega} f \, \mathrm{d}\mu$$

and

$$\mathrm{Var}(f) = \mathrm{E}\left((f - \mathrm{E}(f))^2 \right) = (\mathrm{C}) \int_{\Omega} \left(f - (\mathrm{C}) \int_{\Omega} f \, \mathrm{d}\mu \right)^2 \mathrm{d}\mu$$

provided that the involved integrals exist. Clearly, this happens when f belongs to $\mathcal{M}_b(\Omega, \mathcal{A})$, but we can speak of the expected value of every random variable $f \in L^1(\mu)$ and of the variance of every random variable $f \in L^2(\mu)$. See Sect. 3.5.

Chebyshev's inequality admits the following probabilistic companion:

Theorem 3.1.5 *If $(\Omega, \mathcal{A}, \mu)$ is a normalized capacity space and $f : \Omega \to \overline{\mathbb{R}}$ is a random variable with expected value $\mathrm{E}(f)$ and variance $\mathrm{Var}(f)$, then*

$$\mu\left(\{x : |f(x) - \mathrm{E}(f)| \geq \alpha\}\right) \leq \frac{1}{\alpha^2} \mathrm{Var}(f)$$

for every $\alpha > 0$.

Proof Indeed, according to Theorem 3.1.4,

$$\begin{aligned} \mu\left(\{x : |f(x) - \mathrm{E}(f)| \geq \alpha\}\right) &= \mu\left(\left\{x : |f(x) - \mathrm{E}(f)|^2 \geq \alpha^2\right\}\right) \\ &\leq \frac{1}{\alpha^2}\left((\mathrm{C}) \int_{\Omega} |f(x) - \mathrm{E}(f)|^2 \, \mathrm{d}\mu \right) \\ &= \frac{1}{\alpha^2} \mathrm{Var}(f). \end{aligned}$$

■

The following result represents the analog of *Hölder's inequality* in the context of Choquet integral:

Theorem 3.1.6 *Suppose that $(\Omega, \mathcal{A})$ is a measurable space and $\mu : \mathcal{A} \to \mathbb{R}_+$ is a lower continuous and strongly subadditive capacity. Then for every pair of functions $f, g \in \mathcal{M}_b(\Omega, \mathcal{A})$ and every pair of positive numbers $p, q \in (1, \infty)$ such that $1/p + 1/q = 1$, the following inequality*

$$\left| (\mathrm{C}) \int_{\Omega} fg \, \mathrm{d}\mu \right| \leq \left((\mathrm{C}) \int_{\Omega} |f|^p \, \mathrm{d}\mu \right)^{1/p} \left((\mathrm{C}) \int_{\Omega} |g|^q \, \mathrm{d}\mu \right)^{1/q}$$

holds.

Proof The proof follows closely the classical argument in the case of Lebesgue integral. See Niculescu and Persson [137], Theorem 1.2.2.

According to the modulus inequality for integrals,

$$\left|(\mathrm{C})\int_\Omega fg\,\mathrm{d}\mu\right| \le (\mathrm{C})\int_\Omega |fg|\,\mathrm{d}\mu,$$

so it remains to prove that

$$(\mathrm{C})\int_\Omega |fg|\,\mathrm{d}\mu \le \left((\mathrm{C})\int_\Omega |f|^p\,\mathrm{d}\mu\right)^{1/p}\left((\mathrm{C})\int_\Omega |g|^q\,\mathrm{d}\mu\right)^{1/q}.$$

The last inequality is trivial if at least one of the integrals $A = (\mathrm{C})\int_\Omega |f|^p\,\mathrm{d}\mu$ and $B = (\mathrm{C})\int_\Omega |g|^q\,\mathrm{d}\mu$ is null. Take into account Proposition 3.2.4 (which makes use of the lower continuity of μ). If both these integrals are strictly positive, we can consider the functions

$$u = \frac{|f|}{A^{1/p}} \text{ and } v = \frac{|g|}{B^{1/q}}.$$

According to Young's inequality,

$$uv \le \frac{u^p}{p} + \frac{v^q}{q},$$

and an appeal to the monotonicity and the positive homogeneity of the Choquet integral, combined with its subadditivity (since μ is submodular), leads to the following string of inequalities

$$\begin{aligned}\frac{(\mathrm{C})\int_\Omega |fg|\,\mathrm{d}\mu}{A^{1/p}B^{1/q}} &\le (\mathrm{C})\int_\Omega \left(\frac{|f|^p}{pA} + \frac{|g|^q}{qB}\right)\mathrm{d}\mu \\ &\le \frac{1}{pA}\cdot(\mathrm{C})\int_\Omega |f|^p\,\mathrm{d}\mu + \frac{1}{qB}\cdot(\mathrm{C})\int_\Omega |g|^q\,\mathrm{d}\mu \qquad (3.1)\\ &= \frac{1}{p}+\frac{1}{q} = 1,\end{aligned}$$

which makes clear the conclusion of the theorem. ■

Remark 3.1.7 An inspection of the proof of Theorem 3.1.6 shows that we may replace the condition of submodularity by the comonotonicity of the two functions involved. In this case, the inequalities 3.1 become equalities.

The assertion of Theorem 3.1.6 for $p = q = 2$ is known as the *Cauchy–Bunyakovsky–Schwarz inequality*:

Corollary 3.1.8 *Suppose that $\mu : \mathcal{A} \to \mathbb{R}_+$ is a lower continuous and strongly subadditive capacity and f, g are functions in $\mathcal{M}_b(\Omega, \mathcal{A})$. Then*

$$\left|(\mathrm{C})\int_\Omega fg\,\mathrm{d}\mu\right| \le \left((\mathrm{C})\int_\Omega f^2\,\mathrm{d}\mu\right)^{1/2}\left((\mathrm{C})\int_\Omega g^2\,\mathrm{d}\mu\right)^{1/2}.$$

Another classical inequality that admits an analog in framework of Choquet integral is Minkowski's inequality:

Theorem 3.1.9 (Minkowski's Inequality) *Suppose that $\mu : \mathcal{A} \to \mathbb{R}_+$ is a strongly subadditive capacity and $p \in [1, \infty)$. Then, every pair f, g of functions in $\mathcal{M}_b(\Omega, \mathcal{A})$ whose sum also belongs to $\mathcal{M}_b(\Omega, \mathcal{A})$, verifies the inequality*

$$\left((\mathrm{C})\int_\Omega |f+g|^p\,\mathrm{d}\mu\right)^{1/p} \le \left((\mathrm{C})\int_\Omega |f|^p\,\mathrm{d}\mu\right)^{1/p} + \left((\mathrm{C})\int_\Omega |g|^p\,\mathrm{d}\mu\right)^{1/p}.$$

Proof The case $p = 1$ follows from Theorem 2.6.1. The case $p \in (1, \infty)$ combines the subadditivity of the Choquet integral with Hőlder's inequality as follows:

$$\begin{aligned}
(\mathrm{C})\int_\Omega |f+g|^p\,\mathrm{d}\mu &= (\mathrm{C})\int_\Omega |f+g|^{p-1}\,|f+g|\,\mathrm{d}\mu\\
&\le (\mathrm{C})\int_\Omega |f+g|^{p-1}\,|f|\,\mathrm{d}\mu + (\mathrm{C})\int_\Omega |f+g|^{p-1}\,|g|\,\mathrm{d}\mu\\
&\le \left((\mathrm{C})\int_\Omega |f|^p\,\mathrm{d}\mu\right)^{1/p}\cdot\left((\mathrm{C})\int_\Omega |f+g|^{(p-1)q}\,\mathrm{d}\mu\right)^{1/q}\\
&\quad + \left((\mathrm{C})\int_\Omega |g|^p\,\mathrm{d}\mu\right)^{1/p}\cdot\left((\mathrm{C})\int_\Omega |f+g|^{(p-1)q}\,\mathrm{d}\mu\right)^{1/q}\\
&\le \left[\left((\mathrm{C})\int_\Omega |f|^p\,\mathrm{d}\mu\right)^{1/p} + \left((\mathrm{C})\int_\Omega |g|^p\,\mathrm{d}\mu\right)^{1/p}\right]\left((\mathrm{C})\int_\Omega |f+g|^p\,\mathrm{d}\mu\right)^{1-1/p}
\end{aligned}$$

and the conclusion is now clear. ∎

The extension of the inequalities above to the case of $L^P(\mu)$ spaces is discussed in Corollary 3.5.2.

The next result extends Jensen's inequality to the context of Choquet integral.

Theorem 3.1.10 (Jensen's Inequality for Choquet Integral) *Consider a normalized capacity space* $(\Omega, \mathcal{A}, \mu)$*. Let* $f : \Omega \to \mathbb{R}$ *be a Choquet integrable function taking values in an interval* $[a, b] \subset \mathbb{R}_+$ *and* $\varphi : \mathbb{R} \to \mathbb{R}$ *be a nondecreasing convex function such that* $\varphi \circ f$ *is also Choquet integrable. Then*

$$\varphi\left((\mathrm{C})\int_\Omega f\,\mathrm{d}\mu\right) \le (\mathrm{C})\int_\Omega \varphi\circ f\,\mathrm{d}\mu.$$

If instead φ *is a nondecreasing concave function, then*

$$\varphi\left((\mathrm{C})\int_\Omega f\,\mathrm{d}\mu\right) \ge (\mathrm{C})\int_\Omega \varphi\circ f\,\mathrm{d}\mu.$$

Proof The proof is based on the property of subdifferentiability of convex functions.

Put $\alpha = (\mathrm{C})\int_\Omega f\,\mathrm{d}\mu$. Since φ is a convex function, there is an affine function $h(t) = kt + m$ such that

$$k\alpha + m = \varphi(\alpha) \text{ and } kt + m \le \varphi(t) \text{ for every } t \in \mathbb{R}.$$

Moreover, one can choose $k = f'_+(\alpha)$, which is nonnegative since φ is nondecreasing. See [137], Section 2.1.

Replacing φ by φ plus a suitable positive constant, the conclusion of the theorem remains unchanged (due to the comonotonic additivity), but this operation makes possible to assume that $m \ge 0$. Then

$$\begin{aligned}\varphi\left((\mathrm{C})\int_\Omega f\,\mathrm{d}\mu\right) &= k\cdot\left((\mathrm{C})\int_\Omega f\,\mathrm{d}\mu\right) + m\\ &= (\mathrm{C})\int_\Omega (kf + m)\,\mathrm{d}\mu\\ &\le (\mathrm{C})\int_\Omega \varphi\circ f\,\mathrm{d}\mu.\end{aligned}$$

as desired.

The case of concave functions can be treated similarly. ■

The condition that φ is nondecreasing in Theorem 3.1.10 cannot be dropped as shows the following example due to Zhang, Mesiar and Pap [193]. Consider the case where $\Omega = [0, 1]$, $\mathcal{A} = B([0, 1])$ and $\mu = m^2$, where m is the Lebesgue measure. The function $f(x) = x$ is measurable and the function $\varphi(x) = e^{-x}$ is convex and decreasing. We have

$$(\mathrm{C})\int_\Omega f\,\mathrm{d}\mu = \int_0^1 m^2(\{x\in[0,1] : x \ge t\})\,\mathrm{d}t = \int_0^1 (1-t)^2\,\mathrm{d}t = \frac{1}{3}$$

and

$$\varphi\left((\mathrm{C})\int_{\Omega} f \, \mathrm{d}\mu\right) = e^{-1/3} \approx 0.716\,531\,310\,6.$$

On the other hand,

$$\begin{aligned}(\mathrm{C})\int_{\Omega} \varphi \circ f \, \mathrm{d}\mu &= \int_0^1 m^2\left(\{x \in [0,1] : e^{-x} \geq t\}\right) \mathrm{d}t \\ &= \int_0^{1/e} m^2(\emptyset) \, \mathrm{d}t + \int_{1/e}^1 m^2([0, -\ln t]) \, \mathrm{d}t \\ &= \int_{1/e}^1 \ln^2 t dt = 2 - 5e^{-1} \approx 0.160\,602\,794\,1,\end{aligned}$$

which shows that the statement of Jensen's inequality may fail for the convex and decreasing functions.

Remark 3.1.11 (The Quasi-Sublinearity of Subadditive Capacities) If a capacity $\mu : \mathcal{A} \to \mathbb{R}_+$ is c-subadditive for some constant $c \geq 1$, i.e.,

$$\mu(A \cup B) \leq c(\mu(A) + \mu(B)) \quad \text{for every } A, B \in \mathcal{A},$$

then the Choquet integral associated with it verifies the inequality

$$(\mathrm{C})\int_{\Omega} (f+g) \, \mathrm{d}\mu \leq 2c\left[(\mathrm{C})\int_{\Omega} f \, \mathrm{d}\mu + (\mathrm{C})\int_{\Omega} g \, \mathrm{d}\mu\right]$$

for every functions $f, g \in \mathcal{M}_b^+(\Omega, \mathcal{A})$ whose sum also belongs to $\mathcal{M}_b^+(\Omega, \mathcal{A})$. Moreover, the constant $2c$ should be replaced by 2^{N-1} when N summands are used in the integrand.

Indeed, since

$$\{x : f(x) + g(x) \geq t\} \subset \{x : f(x) \geq t/2\} \cup \{x : g(x) \geq t/2\},$$

it follows that

$$\begin{aligned}(\mathrm{C})\int_{\Omega} (f+g) \, \mathrm{d}\mu &= \int_0^{+\infty} \mu(\{x : f(x) + g(x) \geq t\}) \, \mathrm{d}t \\ &\leq c\int_0^{+\infty} \mu(\{x : f(x) \geq t/2\}) \, \mathrm{d}t + c\int_0^{+\infty} \mu(\{x : g(x) \geq t/2\}) \, \mathrm{d}t \\ &= 2c\left[(\mathrm{C})\int_{\Omega} f \, \mathrm{d}\mu + (\mathrm{C})\int_{\Omega} g \, \mathrm{d}\mu\right].\end{aligned}$$

As was already noticed, if $\mathcal{A}$ is a σ-algebra, then the space $\mathcal{M}_b(\Omega, \mathcal{A})$ is linear. However, even in this case, the Choquet integral may fail to define an L^p norm (due to the lack of strong subadditivity). A substitute is offered by the quasinorms. Indeed, if μ is lower continuous and c-subadditive and $p, q \in (1, \infty)$, then

$$\left|(\mathrm{C})\int_\Omega fg\, \mathrm{d}\mu\right| \le 2c\left((\mathrm{C})\int_\Omega |f|^p\, \mathrm{d}\mu\right)^{1/p}\left((\mathrm{C})\int_\Omega |g|^q\, \mathrm{d}\mu\right)^{1/q},$$

for all $f, g \in \mathcal{M}_b(\Omega, \mathcal{A})$. See Adams [2] and Cerdà et al. [33,34] for more.

3.2 Quasi-Everywhere Properties

As in the case of classical measure theory, it is convenient to have at hand a concept of negligibility associated with a capacity space $(\Omega, \mathcal{A}, \mu)$.

Definition 3.2.1 We say that a property $P(x)$ that depends on the points x of Ω holds μ-quasi-everywhere (equivalently, μ-q.e. or q.e.) if the set of points where $P(x)$ fails is included in a set of zero capacity.

This definition is perfectly suitable for considering functions defined quasi-everywhere, sets or functions equal quasi-everywhere, sequence of functions converging quasi-everywhere, etc.

Notice that

$$A = B \text{ q.e.} \quad \text{if and only if } \chi_A = \chi_B \text{ q.e.}$$

and

$$A_n \nearrow A \text{ q.e.} \quad \text{if and only if } \chi_{A_n} \nearrow \chi_A \text{ q.e.}$$

In general, the union of two sets with zero capacity is not a set of this type. See the case of the capacity C defined on $\mathcal{P}(\{1,2\})$ by the formula $C(A) = 0$ if $A \neq \{1,2\}$ and $C(\{1,2\}) = 1$. The same example shows that the equality $\chi_A = \chi_B$ q.e. does not necessarily imply the equality of their capacities.

If μ is a subadditive capacity, then the finite union of zero capacity sets is a set of zero capacity. Under the same circumstances, $A = B$ q.e. implies $\mu(A) = \mu(B)$. See Proposition 3.2.3.

Remark 3.2.2 If μ is a lower continuous subadditive capacity defined on a σ-algebra, then the countable union of zero capacity sets is a set of zero capacity. This allows us to simplify the proof of results concerning properties that hold almost everywhere by reducing the considerations to the case when those properties hold everywhere.

Proposition 3.2.3 *Suppose that μ is a subadditive capacity. If $f, g \in \mathcal{M}(\Omega, \mathcal{A})$, $f = g$ q.e. and f is Choquet integrable, then the function g is also integrable and*

$$(\mathrm{C}) \int_\Omega f \, \mathrm{d}\mu = (\mathrm{C}) \int_\Omega g \, \mathrm{d}\mu.$$

In particular, if $A, B \in \mathcal{A}$ and $A = B$ q.e., then $\mu(A) = \mu(B)$.

Proof Indeed, denote $A = \{x : f(x) \neq g(x)\}$. Then there exists a set E of zero capacity that includes A, which implies that for every $t \in \mathbb{R}$, we get

$$\begin{aligned}
\mu(\{x : f(x) > t\}) &= \mu\left((\{x : f(x) > t\} \cap \complement E) \cup (\{x : f(x) > t\} \cap E)\right) \\
&\leq \mu\left((\{x : g(x) > t\} \cap \complement E)) \cup E\right) \\
&\leq \mu\left(\{x : g(x) > t\} \cap \complement E)\right) + \mu(E) \\
&= \mu\left(\{x : g(x) > t\} \cap \complement E)\right) \\
&\leq \mu(\{x : g(x) > t\}).
\end{aligned}$$

Noticing that the roles of f and g can be interchanged, we conclude that actually the equality holds. Now the assertion of the proposition is clear. ■

Proposition 3.2.4 *Suppose that $(\Omega, \mathcal{A}, \mu)$ is a capacity space and $f : \Omega \to \mathbb{R}_+ \cup \{\infty\}$ is an $\mathcal{A}$-measurable function.*

(i) *If $f = 0$ q.e., then* $(\mathrm{C}) \int_\Omega f \, \mathrm{d}\mu = 0$.
(ii) *If μ is lower continuous and* $(\mathrm{C}) \int_\Omega f \, \mathrm{d}\mu = 0$,, *then $f = 0$ q.e.*
(iii) *If μ is lower continuous and f is μ-integrable, then f is finite q.e.*

Proof

(i) According to our hypothesis, $\mu(\{x : f(x) > 0\}) = 0$. Due to the inclusion

$$\{x : f(x) > t\} \subset \{x : f(x) > 0\} \quad \text{for every } t > 0$$

and the monotonicity of the capacity μ, it follows that $\mu(\{x : f(x) > t\}) = 0$ for every $t > 0$. Hence

$$(\mathrm{C}) \int_\Omega f \, \mathrm{d}\mu = \int_0^\infty \mu(\{x : f(x) > t\}) \, \mathrm{d}t = 0.$$

(ii) Notice first that

$$\text{(C)} \int_{\Omega} f \, d\mu = 0 \text{ implies } \mu(\{x : f(x) \geq t\}) = 0 \text{ for every } t > 0.$$

Indeed, if $\mu(\{x : f(x) \geq t_0\}) = \alpha > 0$ for some $t_0 > 0$,, then

$$\mu(\{x : f(x) \geq t\}) \geq \alpha \quad \text{for all } t \in (0, t_0]$$

since μ is nondecreasing. This would imply

$$0 = \text{(C)} \int_A f \, d\mu = \int_0^{\infty} \mu(\{x : f(x) > t\}) \, dt \geq \int_0^{t_0} \mu(\{x : f(x) > t\}) \, dt \geq \alpha t_0,$$

which is a contradiction. Therefore $\mu(\{x : f(x) \geq t\}) = 0$ for every $t > 0$. The fact that

$$\{x : f(x) > 0\} = \bigcup_{n=1}^{\infty} \{x : f(x) \geq 1/n\}$$

shows that $\{x : f(x) > 0\}$ is the union of a nondecreasing sequence of sets in $\mathcal{A}$, so by the lower continuity of μ, we conclude that

$$\mu(\{x : f(x) > 0\}) = \lim_{n \to \infty} \mu(\{x : f(x) \geq 1/n\}) = 0.$$

(iii) Notice first that $\{x : f(x) = \infty\} = \bigcap_{\alpha > 0} \{x : f(x) \geq \alpha\}$. Since the integral of f is finite and

$$\begin{aligned}
\text{(C)} \int_{\Omega} f \, d\mu &= \int_0^{\infty} \mu(\{x : f(x) > t\}) \, dt \\
&\geq \int_0^{\alpha} \mu(\{x : f(x) > t\}) \, dt \\
&\geq \alpha \mu(\{x : f(x) > \alpha\})
\end{aligned}$$

for every $\alpha > 0$, it follows that

$$0 \leq \mu(\{x : f(x) = \infty\}) \leq \frac{1}{\alpha} \text{(C)} \int_{\Omega} f \, d\mu \to 0 \quad \text{as } \alpha \to \infty,$$

and the proof is done.

■

The following consequence of Proposition 3.2.4 plays a role in the construction of the capacitary spaces.

Corollary 3.2.5 *Suppose that μ is a lower continuous and subadditive capacity defined on a σ-algebra Σ of subsets of Ω and $f, g \in \mathcal{M}(\Omega, \Sigma)$. Then*

$$(\mathrm{C})\int_\Omega |f-g|\,\mathrm{d}\mu = 0 \text{ if and only if } f = g \text{ q.e.}$$

Handling measurable functions is much easier in the case of complete capacity spaces.

Definition 3.2.6 A capacity space (Ω, Σ, μ) is called complete if

$$S \subset N,\ N \in \Sigma \text{ and } \mu(N) = 0 \text{ implies } S \in \Sigma.$$

So is the case when $\Sigma = \mathcal{P}(\Omega)$. Otherwise, assuming that Σ is a σ-algebra, one can consider the *completion* $(\Omega, \Sigma_\mu, \widetilde{\mu})$ of the capacity space (Ω, Σ, μ), where Σ_μ is the σ-algebra defined as

$$\Sigma_\mu = \{A \cup S : A \in \Sigma \text{ and } S \text{ is a subset of some set } N \in \Sigma \text{ with } \mu(N) = 0\}$$

and $\widetilde{\mu}$ is the capacity defined on Σ_μ by the formula

$$\widetilde{\mu}(A \cup S) = \mu(A).$$

Notice that $\Sigma \subset \Sigma_\mu$ and $\widetilde{\mu}|_\Sigma = \mu$.

Usually, Σ_μ is called the σ-algebra of *μ-measurable sets* and the functions $f : \Omega \to \overline{\mathbb{R}}$ measurable with respect to $\widetilde{\Sigma}$, are called *μ-measurable functions*. These notions constitute a convenient tool when dealing with capacity spaces that are not necessarily complete.

The permanence properties related to the operation of completion are summarized as follows.

Proposition 3.2.7 *Suppose that (Ω, Σ) is a measurable space and $(\Omega, \Sigma_\mu, \widetilde{\mu})$ is the completion of the capacity space (Ω, Σ, μ).*

(i) *If μ is subadditive (countably subadditive or strongly subadditive),, then so is $\widetilde{\mu}$.*
(ii) *If μ is lower continuous and countably subadditive, then $\widetilde{\mu}$ is lower continuous.*

The details are straightforward.

A consequence of Propositions 3.2.4 and 3.2.7 is the possibility to replace the extended real-valued measurable functions that take infinite values only on sets of zero capacity by real-valued measurable functions.

3.3 Sequences of Measurable/Integrable Functions

The quasi-everywhere convergence of the sequences of measurable functions makes the objective of the following result.

Theorem 3.3.1 *Suppose that* (Ω, Σ) *is a measurable space and* μ *is a complete capacity defined on* Σ*. If* $(f_n)_n$ *is a sequence of* Σ*-measurable functions (possibly taking values in* $\overline{\mathbb{R}}$*) and* g *is a function defined on* Ω *such that*

$$\lim_{n\to\infty} f_n(x) = g(x) \quad q.e.,$$

then g *is a measurable function.*

Proof Suppose first that $\lim_{n\to\infty} f_n(x) = g(x)$ for every $x \in \Omega$. We have to prove that all the sets $\{x : g(x) > t\}$, with $t \in \mathbb{R}$, are measurable. For this, consider the measurable sets

$$A_k^{(m)} = \left\{x : f_k(x) > t + \frac{1}{m}\right\} \quad (k, m \in \mathbb{N})$$

and notice the equality

$$\{x : g(x) > t\} = \bigcup\nolimits_{n,m=1}^{\infty} \left(\bigcap\nolimits_{k=n}^{\infty} A_k^{(m)}\right). \tag{3.2}$$

This can be proved by inclusions. If $x_0 \in \{x : g(x) > t\}$,, then $g(x_0) > t$, and one can choose a natural number m such that $g(x_0) > t + 1/m$. Since $f_k(x_0) \to g(x_0)$,, there is an $n \in \mathbb{N}$ such that $f_k(x_0) > t + 1/m$. for all $k \geq n$. Hence,

$$\{x : g(x) > t\} \subset \bigcup\nolimits_{n,m=1}^{\infty} \left(\bigcap\nolimits_{k=n}^{\infty} A_k^{(m)}\right).$$

For the reverse inclusion, let

$$x_0 \in \bigcup\nolimits_{n,m=1}^{\infty} \left(\bigcap\nolimits_{k=n}^{\infty} A_k^{(m)}\right).$$

Then $x_0 \in \bigcap_{k=n}^{\infty} A_k^{(m)}$ for some $m, n \in \mathbb{N}$,, which yields

$$f_k(x_0) > t + \frac{1}{m} \text{ for all } k \geq n.$$

Passing to the limit as $k \to \infty$, we infer that $g(x_0) \geq t + \frac{1}{m} > t$, i.e., x_0 belongs to $\{x : g(x) > t\}$. This ends the proof of the equality (3.2) and also of the measurability of g since t was arbitrarily fixed.

If $\lim_{n\to\infty} f_n(x) = g(x)$ q.e., let us denote by E the exceptional set of all points x such that $\lim_{n\to\infty} f_n(x) \neq g(x)$. Then E is a subset of a set of capacity zero, and due to the fact that μ is complete, it follows that $E \in \Sigma$ (and $\mu(E) = 0$). Then $g|_{\Omega\setminus E}$ is measurable by the first part of our reasoning and $g|_E$ is measurable because $\mu(E) = 0$ and μ is complete. Then, for every Borel subset B of $\overline{\mathbb{R}}$,

$$g^{-1}(B) = \left(g|_{\Omega\setminus E}\right)^{-1}(B) \cup (g|_E)^{-1}(B) \in \Sigma,$$

which ends the proof of the measurability of g. ■

Corollary 3.3.2 *Under the hypotheses of Theorem* 3.3.1, *if $f : \Omega \to \overline{\mathbb{R}}$ is a measurable function and $g : \Omega \to \overline{\mathbb{R}}$ is a function equal q.e. to f,, then g is measurable too.*

Let us consider yet another important type of convergence of measurable functions.

Definition 3.3.3 A sequence $(f_n)_n$ of measurable functions is convergent in capacity to a measurable function f if for every $\varepsilon > 0$,

$$\mu\left(\{x : |f_n(x) - f(x)| \geq \varepsilon\}\right) \to 0 \quad \text{as } n \to \infty.$$

The sequence $(f_n)_n$ is Cauchy in capacity if for every $\varepsilon > 0$,

$$\mu\left(\left\{x : \left|f_i(x) - f_j(x)\right| \geq \varepsilon\right\}\right) \to 0 \quad \text{as } i, j \to \infty.$$

The convergence in capacity does not imply the q.e. convergence. Indeed, this does not hold even in the case of the Lebesgue measure. See Bogachev [25], Example 2.2.4, p. 112. However, in the case of finite measure spaces, the almost everywhere convergence implies convergence in measure. See [25], Theorem 2.2.3, p. 111.

Lemma 3.3.4 *Assume that (Ω, Σ, μ) is a capacity space, where μ is a strongly subadditive, lower continuous, and complete capacity. A sequence $(f_n)_n$ of measurable functions is convergent in capacity to a measurable function f if and only if it is a Cauchy sequence in capacity. In the later case, the sequence has a subsequence that is q.e. convergent to f.*

Proof If $(f_n)_n$ is a Cauchy sequence in capacity, then there exist numbers $n_k \in \mathbb{N}$ such that

$$\mu\left(\left\{x : \left|f_i(x) - f_j(x)\right| \geq 1/2^k\right\}\right) < 1/2^k \text{ for all } i, j \geq n_k > n_{k-1}.$$

Put

$$A_k = \left\{x : \left|f_{n_k}(x) - f_{n_{k+1}}(x)\right| \geq 1/2^k\right\} \text{ and } F_m = \bigcup\nolimits_{k\geq m} A_k.$$

If $j \geq i \geq m$, then $\left|f_{n_i}(x) - f_{n_j}(x)\right| < 1/2^{m-1}$ on $\Omega \setminus F_m$. As a consequence, $(f_n)_n$ is uniformly Cauchy on $\Omega \setminus F_m$, and it is simply convergent to a function f on $E = \bigcup_m (\Omega \setminus F_m)$. Since

$$\mu(\Omega \setminus E) = \mu\left(\complement\left(\bigcup\nolimits_m (\Omega \setminus F_m)\right)\right) = \mu\left(\bigcap\nolimits_m F_m\right) \leq \mu\left(F_{m_0}\right)$$

for every index m_0 and

$$\begin{aligned}\mu\left(F_{m_0}\right) &= \mu\left(\bigcup\nolimits_{k \geq m_0} A_k\right) \\ &\leq \sum_{k=m_0}^{\infty} \mu\left(A_k\right) < \sum_{k=m_0}^{\infty} \frac{1}{2^k} = \frac{1}{2^{m_0-1}} \to 0 \text{ as } m_0 \to \infty,\end{aligned}$$

because μ is countably subadditive, it follows that $\mu(\Omega \setminus E) = 0$. By Theorem 3.3.1 it follows that f is a measurable function. On the other hand, according to Theorem 3.4.1 (ii) in the next section,

$$\begin{aligned}\mu\left(\left\{x : \left|f_{n_k}(x) - f(x)\right| > \delta\right\}\right) &= \mu\left(\left\{x : \lim_{j\to\infty} \left|f_{n_k}(x) - f_{n_j}(x)\right| > \delta\right\}\right) \\ &\leq \lim_{j\to\infty} \mu\left(\left\{x : \left|f_{n_k}(x) - f_{n_j}(x)\right| > \delta\right\}\right) < \varepsilon.\end{aligned}$$

Since $(f_n)_n$ is a Cauchy sequence in capacity, which has a subsequence that is convergent in capacity to f, it follows that the entire sequence $(f_n)_n$ converges to f in capacity. ■

The next results refer to the behavior of sequences of integrable functions.

Theorem 3.3.5 (The Monotone Convergence Theorem) *Suppose that (Ω, Σ) is a measurable space and $\mu : \Sigma \to \mathbb{R}_+ \cup \{\infty\}$ is a lower continuous capacity. Then, given an increasing sequence of nonnegative measurable functions $f_n : \Omega \to \mathbb{R}_+ \cup \{\infty\}$, the function $\sup_{n\in\mathbb{N}} f_n$ defined by the formula*

$$\left(\sup_{n\in\mathbb{N}} f_n\right)(x) = \sup_{n\in\mathbb{N}} f_n(x), \quad x \in \Omega$$

is measurable and

$$(\mathrm{C})\int_{\Omega} \sup_{n\in\mathbb{N}} f_n \, d\mu = \sup_{n\in\mathbb{N}}\left((\mathrm{C})\int_{\Omega} f_n \, d\mu\right) = \lim_{n\to\infty} (\mathrm{C})\int_{\Omega} f_n \, d\mu.$$

Proof Since Σ is a σ-algebra, the measurability of the function $f = \sup_{n\in\mathbb{N}} f_n$ reduces to the fact that all upper level sets $\{x : f(x) > \alpha\}$ are measurable whenever $\alpha \in \mathbb{R}_+$. Since the sequence $(f_n)_n$ is increasing, it follows that

$$\{x : f_n(x) > \alpha\} \subset \{x : f_{n+1}(x) > \alpha\} \subset \{x : f(x) > \alpha\}$$

and thus

$$\bigcup\nolimits_{n=1}^{\infty} \{x : f_n(x) > \alpha\} \subset \{x : f(x) > \alpha\}.$$

Actually equality holds because $f_n(x) \to f(x)$ for every x. Thus, $\{x : f(x) > \alpha\}$ is the countable union of an increasing sequence of measurable sets whenever $\alpha \in \mathbb{R}_+$, which shows that f is measurable.

Since μ is lower continuous,

$$\mu(\{x : f_n(x) > \alpha\}) \nearrow \mu(\{x : f(x) > \alpha\}) \quad \text{as } n \to \infty,$$

so, by the classical Theorem of Monotone convergence from measure theory (see Theorem 1.1.12), we infer that

$$(\mathrm{C})\int_{\Omega} f_n \,\mathrm{d}\mu = \int_0^{\infty} \mu(\{x : f_n(x) > t\})\,\mathrm{d}t$$

converges to

$$\int_0^{\infty} \mu(\{x : f(x) > t\})\,\mathrm{d}t = (\mathrm{C})\int_{\Omega} f \,\mathrm{d}\mu$$

as $n \to \infty$. ■

From Theorem 3.3.5, one can deduce the following Choquet integral variant of Fatou's Lemma.

Lemma 3.3.6 (Fatou's Lemma) *Given a complete capacity space* (Ω, Σ, μ), *let* $(f_n)_n$ *be a sequence of measurable functions taking values in* $\mathbb{R}_+ \cup \{\infty\}$. *Then the function* $f(x) = \liminf_{n\to\infty} f_n(x)$ *is also measurable and*

$$(\mathrm{C})\int_{\Omega} \liminf_{n\to\infty} f_n \,\mathrm{d}\mu \le \liminf_{n\to\infty}\left((\mathrm{C})\int_{\Omega} f_n \,\mathrm{d}\mu\right).$$

Here $\liminf_{n\to\infty}$ *denotes the* limit inferior, *i.e.*,

$$\liminf_{n\to\infty} f_n = \sup_{n\ge 1} \inf_{k\ge n} f_k.$$

Proof Put $g_n(x) = \inf_{k \geq n} f_k(x)$. Then the functions g_n are nonnegative and measurable (due to Theorem 3.3.1) and

$$\left(\liminf_{n \to \infty} f_n\right)(x) = \sup_n g_n(x).$$

Since the sequence $(g_n)_n$ is increasing, we infer from the Monotone Convergence Theorem that

$$(\mathrm{C}) \int_\Omega \liminf_{n \to \infty} f_n \, \mathrm{d}\mu = \lim_{n \to \infty} \left((\mathrm{C}) \int_\Omega g_n \, \mathrm{d}\mu \right).$$

Taking into account that $g_n \leq f_k$ for all $k \geq n$,, we obtain the inequality

$$(\mathrm{C}) \int_\Omega g_n \, \mathrm{d}\mu \leq \inf_{k \geq n} \left((\mathrm{C}) \int_\Omega f_n \, \mathrm{d}\mu \right),$$

which implies that

$$\lim_{n \to \infty} (\mathrm{C}) \int_\Omega g_n \, \mathrm{d}\mu \leq \lim_{n \to \infty} \inf_{k \geq n} \left((\mathrm{C}) \int_\Omega f_n \, \mathrm{d}\mu \right) = \liminf_{n \to \infty} \left((\mathrm{C}) \int_\Omega f_n \, \mathrm{d}\mu \right).$$

■

It is worth mentioning that Fatou's Lemma allows us to establish a result that extends the Monotone Convergence Theorem:

Theorem 3.3.7 (J.F. Feinstein [61]) *Suppose that (Ω, Σ) is a measurable space and $\mu : \Sigma \to \mathbb{R}_+ \cup \{\infty\}$ is a lower continuous, subadditive, and complete capacity. If $f_n : \Omega \to \mathbb{R}_+ \cup \{\infty\}$ is a sequence of measurable functions converging almost everywhere to a function f so that $f_n \leq f$ q.e. for all n, then*

$$\lim_{n \to \infty} \left((\mathrm{C}) \int_\Omega f_n \, \mathrm{d}\mu \right) = (\mathrm{C}) \int_\Omega f \, \mathrm{d}\mu.$$

Proof Indeed, taking into account Remark 3.2.2 and Corollary 3.3.2, we may assume that $f_n \leq f$ q.e. for all $n \geq 1$. Then, according to Fatou's Lemma 3.3.6,

$$\begin{aligned} (\mathrm{C}) \int_\Omega f \, \mathrm{d}\mu &= (\mathrm{C}) \int_\Omega \liminf_{n \to \infty} f_n \, \mathrm{d}\mu \leq \liminf_{n \to \infty} \left((\mathrm{C}) \int_\Omega f_n \, \mathrm{d}\mu \right) \\ &\leq \limsup_{n \to \infty} \left((\mathrm{C}) \int_\Omega f_n \, \mathrm{d}\mu \right) \leq (\mathrm{C}) \int_\Omega f \, \mathrm{d}\mu, \end{aligned}$$

and the proof is complete. ■

3.4 The Decreasing Rearrangements Associated with a Capacity Space

The decreasing rearrangement of a measurable function can be introduced in the context of capacity spaces $(\Omega, \mathcal{A}, \mu)$ as a process connecting them with the Lebesgue measurable spaces $([0, \mu(\Omega)], \mathcal{B}([0, \mu(\Omega)]), m)$.

Precisely, the decreasing rearrangement of a set $A \in \mathcal{A}$ is the interval $A^* = [0, \mu(A)]$. The fact that

$$\mu(A) = m(A^*)$$

represents the conservation of the measure under rearrangement.

The decreasing rearrangement of a function $f \in \mathcal{M}(\Omega, \mathcal{A})$ is the function f^* defined on the interval $[0, \mu(\Omega)]$ by the formula

$$\begin{aligned} f^*(t) &= \inf\{s \in [0, \infty] : \mu(\{x : |f(x)| > s\}) \leq t\} \\ &= \sup\{s \in [0, \infty] : 0 \leq t \leq \mu(\{x : |f(x)| > s\})\} \\ &= \sup\left\{s \in [0, \infty] : t \in \{x : |f(x)| > s\}^*\right\}; \end{aligned}$$

we make the convention to define the infimum of an empty set as 0.

Notice that f^* is the right-continuous nonincreasing function from $[0, \mu(\Omega)]$ into $[0, \infty)$, which is equimeasurable with f, i.e.

$$\mu(\{x : |f(x)| > s\}) = m\left(\left\{t : f^*(t) > s\right\}\right).$$

Since $\{s : \mu(\{x : |f(x)| > s\}) \leq t\}$ is the interval $[f^*(t), \infty]$, it follows that

$$f^*(t) = \int_0^\infty \chi_{[0,\mu(\{|f|>s\}))}(t)\, \mathrm{d}s. \tag{3.3}$$

Some authors (e.g., Cerdà, Martín and Silvestre [34] and Ponce and Spector [151] use the notation A^*_μ for A^* (and f^*_μ for f^*) in order to specify the capacity under attention.

Some immediate properties of the decreasing rearrangements are as follows:

(DR1) If $A \in \mathcal{A}$,, then $\chi^*_A = \chi_{A^*}$.
(DR2) If $f, g \in \mathcal{M}^+(\Omega, \mathcal{A})$ and $f \leq g$,, then $f^* \leq g^*$.
(DR3) By Fubini's theorem and formula (3.3), if $f \in \mathcal{M}^+(\Omega, \mathcal{A})$,, then

$$(C)\int_\Omega f\, \mathrm{d}\mu = \int_0^\infty f^*(t)\, \mathrm{d}t.$$

(DR4) If $f = \sum_{i=1}^{n} \alpha_i \chi_{A_i}$ where $A_i \cap A_j = \emptyset$ if $i \neq j$ and $\alpha_1 \geq \alpha_2 \geq \cdots \geq \alpha_n \geq \alpha_{n+1} = 0$, then

$$f^* = \sum_{i=1}^{n} (\alpha_i - \alpha_{i+1})\, \chi_{[0,\mu(A_1 \cup \cdots \cup A_k)]}.$$

The following extension of Fatou's Lemma is due to Cerdà et al. [34]:

Theorem 3.4.1 (The Extension of Fatou's Lemma) *Suppose that* (Ω, Σ) *is a measurable space and* $\mu : \Sigma \to \mathbb{R}_+$ *is a subadditive complete capacity. Then the following properties are equivalent:*

(i) μ *is a lower continuous capacity;*
(ii) *if* $A_n \nearrow A$ *q.e. in* Σ, *then* $\mu(A_n) \to \mu(A)$;
(iii) $f \leq \liminf\limits_{n\to\infty} f_n$ *in* $\mathcal{M}^+(\Omega, \Sigma)$ *implies* $f^* \leq \liminf\limits_{n\to\infty} f_n^*$;
(iv) $(C)\int_\Omega \liminf\limits_{n\to\infty} f_n \,\mathrm{d}\mu \leq \liminf\limits_{n\to\infty} \left((C)\int_\Omega f_n \,\mathrm{d}\mu\right)$ *for every sequence* $(f_n)_n$ *of functions in* $\mathcal{M}^+(\Omega, \Sigma)$;
(v) $f_n \nearrow f$ *q.e. in* $\mathcal{M}^+(\Omega, \Sigma)$ *implies* $f_n^* \nearrow f^*$.

Proof The equivalence (i) $\Leftrightarrow$ (ii) is a consequence of Proposition 3.2.3.

The assertion (ii) follows from (iv) by choosing $f_n = \chi_{A_n}$.

(iv) $\Rightarrow$ (ii) Suppose that $(A_n)_n$ is a nondecreasing sequence in $\mathcal{A}$ such that $A = \bigcup_{n\geq 1} A_n \in \mathcal{A}$. Then $\liminf\limits_{n\to\infty} \chi_{A_n} = \chi_A$ and, according to (iv), $\mu(A)$ does not exceed $\lim\limits_{n\to\infty} \mu(A_n)$.

(iv) follows from (iii), the property (DR3) and the classical Fatou's lemma.

(ii) $\Rightarrow$ (iii) Suppose that $f \leq \liminf\limits_{n\to\infty} f_n$ in $\mathcal{M}^+(\Omega, \Sigma)$. Then, for every $t > 0$,

$$\{x : f(x) > t\} \subset \bigcup_{n\geq 1} \bigcap_{k\geq n} \{x : f_k(x) > t\}$$

which implies that

$$\begin{aligned}\mu(\{x : f(x) > t\}) &\leq \lim_{n\to\infty} \mu\left(\bigcap_{k\geq n} \{x : f_k(x) > t\}\right)\\ &\leq \liminf_{n\to\infty} \mu(\{x : f_k(x) > t\}).\end{aligned}$$

Hence,

$$\chi_{[0,\mu(\{x:f(x)>t\})]} \leq \liminf_{n\to\infty} \chi_{[0,\mu(\{x:f_n(x)>t\})]}$$

and according to (iii), we infer that

$$\begin{aligned} f^*(t) = f^*(t) &= \int_0^\infty \chi_{[0,\mu(\{x:f(x)>s\})}(t)\,\mathrm{d}s \\ &\leq \liminf_{n\to\infty} \int_0^\infty \chi_{[0,\mu(\{x:f_n(x)>s\})}(t)\,\mathrm{d}s \\ &= \liminf_{n\to\infty} f_n^*(t). \end{aligned}$$

(v) $\Rightarrow$ (ii) results by taking $f_n = \chi_{A_n}$ and $f = \chi_A$.

(ii) $\Rightarrow$ (iii) Suppose that $0 \leq f_n \nearrow f$ in $\mathcal{M}(\Omega, \Sigma)$. Then $f_n^* \leq f^*$ for all n, which implies that $\lim_{n\to\infty} f_n^*(x) \leq f^*(x)$ for all x. According to (i), $\mu(\{x : |f(x)| > t\}) = \lim_{n\to\infty} \mu(\{x : |f_n(x)| > t\})$ and thus

$$f^*(x) = \int_0^\infty \chi_{[0,\mu(\{|f|>t\})}(x)\,\mathrm{d}t = \int_0^\infty \lim_{n\to\infty} \chi_{[0,\mu(\{|f_n|>t\})}(x)\,\mathrm{d}t \leq \lim_{n\to\infty} f_n^*(x).$$

The proof is complete. ∎

The extension of Fatou's lemma will be used in the next section to prove the completion of the capacitary spaces.

Based on the decreasing rearrangements, Ridaoui and Grabisch [157] were able to extend the results of Sugeno from Sect. 2.7, to calculate the Choquet integrals of continuous and nonnegative functions, which are not necessarily monotonic. Just to offer an idea, let us consider the case where $f : [0, t] \to \mathbb{R}_+$ is nondecreasing on $[0, \tau_0]$, and nonincreasing on $[\tau_0, t]$ and $f(t) \geq f(0)$. Put $\overline{f} = f(\tau_0) = \max\{f(\tau) : 0 \leq \tau \leq t\}$ and consider the function

$$m_f : [f(0), \overline{f}] \to [0, t], \quad m_f(r) = m(\{\tau;\ f(\tau) \geq r\}).$$

The function m_f is continuous, and nonincreasing on $[f(0), \overline{f}]$ and is related to the nondecreasing rearrangement $f^* : [0, t] \to [f(0), \overline{f}]$ of f by the formula

$$f^*(\tau) = m_f^{-1}(t - \tau). \tag{3.4}$$

Due to the equimeasurability of f and f^*,

$$m(\{\tau : f(\tau) \geq r\}) = m\left(\left\{\tau : f^*(\tau) \geq r\right\}\right) \quad \text{for every } r \in [f(0), \overline{f}],$$

the following equality

$$(\mathrm{C}) \int_{[0,t]} f\,\mathrm{d}m_\varphi = (\mathrm{C}) \int_{[0,t]} f^*\,\mathrm{d}m_\varphi$$

holds. Thus, taking into account Lemma 2.7.1, we conclude that

$$(\mathrm{C})\int_{[0,t]} f\,\mathrm{d}m_\varphi = \int_0^t \varphi'(t-\tau)f^*(\tau)\,\mathrm{d}\tau.$$

Applications of this last formula are available in the paper of Ridaoui and Grabisch [157].

3.5 The Capacitary Function Spaces $L^p(\mu)$

Throughout this section, $\mu : \Sigma \to \mathbb{R}_+$ will denote a lower continuous, strongly subadditive finite, and complete capacity defined on a σ-algebra Σ of subsets of a total set Ω. Under these assumptions, the space $\mathcal{M}(\Omega, \Sigma)$ is an order σ-complete vector lattice, and the Choquet integral is a sublinear functional. See respectively Lemma 2.3.5 and Theorem 2.6.1.

For $p \in [1, \infty)$, one denotes by $\mathcal{L}^p(\mu)$ the vector space of real-valued measurable functions $f \in \mathcal{M}(\Omega, \Sigma)$ such that

$$\|f\|_{L^p(\mu)} = \left((\mathrm{C})\int_\Omega |f|^p\,\mathrm{d}\mu\right)^{1/p} < \infty.$$

Notice that

$$\begin{aligned}\|f\|_{L^p(\mu)} &= \left(\left\||f|^p\right\|_{L^1(\mu)}\right)^{1/p}\\ &= \left(p\int_0^\infty t^{p-1}\mu\left(\{x : |f(x)| > t\}\right)\mathrm{d}t\right)^{1/p},\end{aligned}$$

via a change of variable in the Lebesgue integral.

The fact that the function $\|\cdot\|_{L^p(\mu)}$ is a seminorm on the space $\mathcal{L}^p(\mu)$ follows from Theorem 3.1.9 (the generalized Minkowski inequality) and the following result.

Theorem 3.5.1 *Suppose that (Ω, μ) is a measurable space and $\mu : \Sigma \to \mathbb{R}_+$ is a lower continuous, strongly subadditive, and complete capacity. Then for every function $f \in \mathcal{L}^p(\mu)$ with $p \in [1, \infty)$ there exists a sequence of bounded measurable functions $f_n \in \mathcal{M}_b(\Omega, \Sigma)$ such that*

$$\lim_{n\to\infty}(\mathrm{C})\int_\Omega |f - f_n|^p\,\mathrm{d}\mu = 0.$$

Proof For each $n \in \mathbb{N}$, one considers the truncation of f at height n,, i.e., the function $f_n = (f \vee -n) \wedge n$, which belongs to $\mathcal{M}_b(\Omega, \Sigma)$ because this set is a lattice. Since

$$f_n(x) = \begin{cases} -n & \text{if } f(x) < -n \\ f(x) & \text{if } f(x) \in [-n, n] \\ n & \text{if } f(x) > n, \end{cases}$$

we have

$$\begin{aligned} \text{(C)} \int_\Omega |f - f_n|^p \, d\mu &= p \int_0^\infty t^{p-1} \mu(\{x : |f(x) - f_n(x)| > t\}) \, dt \\ &= p \int_0^\infty t^{p-1} \mu(\{x : |f(x)| \geq t + n\}) \, dt \\ &= p \int_n^\infty t^{p-1} \mu(\{x : |f(x)| \geq t\}) \, dt \to 0 \end{aligned}$$

as $n \to \infty$ and the proof is done. ■

Corollary 3.5.2 *Under the hypotheses of Theorem* 3.5.1, *the generalized Minkowski inequality stated by Theorem* 3.1.9 *extends verbatim to the framework of spaces* $\mathcal{L}^p(\mu)$ *with* $p \in [1, \infty)$.

The same is true in the case of the modulus inequality and of the Hőlder inequality.

Since $\|\cdot\|_{L^p(\mu)}$ is only a seminorm, we introduce an equivalence relation among the elements in $\mathcal{L}^p(\mu)$ by denoting $f \sim g$ whenever $\|f - g\|_{L^p(\mu)} = 0$, equivalently, $f = g$ q.e. This is clear for $p = \infty$. For $p \in [1, \infty)$, see Corollary 3.2.5.

The space $L^p(\mu)$ is defined as the space $\mathcal{L}^p(\mu)/\sim$, of equivalence classes $[f]$ of functions f in $\mathcal{L}^p(\mu)$, endowed with the quotient algebraic structure defined by

$$\begin{aligned} [f] + [g] &= [f + g] \\ \alpha[f] &= [\alpha f], \end{aligned}$$

the ordering

$$[f] \leq [g] \quad \text{if } f \leq g \text{ q.e.},$$

and the norm

$$\|[f]\|_{L^p(\mu)} = \left(\text{(C)} \int_\Omega |f|^p \, d\mu \right)^{1/p}.$$

As in the case of Lebesgue measure, we will refer to $L^p(\mu)$ as *the space of all* pth *μ-integrable functions*, which is formally incorrect, but convenient. By this, the functions equal almost everywhere are regarded as the same element.

Remark 3.5.3 (The Extension of Basic Inequalities to the Framework of Capacitary Spaces) The assertions of Corollary 3.5.2 also work in the context of spaces $L^p(\mu)$ with $p \in [1, \infty)$.

Theorem 3.5.4 (The Completeness of the L^p-Spaces) *If $\mu : \Sigma \to \mathbb{R}_+$ is a strongly subadditive, lower continuous, and complete capacity defined on a σ-algebra Σ, then all the spaces $L^p(\mu)$ with $p \in [1, \infty)$ are Banach spaces.*

Proof Let $(f_n)_n$ be a Cauchy sequence in $L^p(\mu)$. Then we may choose positive integers $n_1 < n_2 < n_3 < \cdots$ such that

$$\|f_m - f_n\|_{L^p(\mu)} < \left(\frac{1}{2^j}\right)^{p+1} \quad \text{for all } m, n \geq n_j.$$

This implies

$$p \int_0^{1/2^j} t^{p-1} \mu\left(\left\{x : \left|f_{n_j}(x) - f_{n_{j+1}}(x)\right| > 1/2^j\right\}\right) \mathrm{d}t < \left(\frac{1}{2^j}\right)^{p+1},$$

whence

$$\mu\left(\left\{x : \left|f_{n_j}(x) - f_{n_{j+1}}(x)\right| > 1/2^j\right\}\right) \mathrm{d}t < \frac{1}{2^j}.$$

Let $F_j = \left\{x : \left|f_{n_j}(x) - f_{n_{j+1}}(x)\right| > 1/2^j\right\}$ and $E_m = \bigcup_{j \geq m} F_j$. Then for every $x \in \Omega \setminus E_m$,

$$\sum\nolimits_{j \geq m} \left|f_{n_j}(x) - f_{n_{j+1}}(x)\right| \leq \sum\nolimits_{j \geq m} \frac{1}{2^j} = \frac{1}{2^{m-1}}$$

and we may consider the function f defined on $\Omega \setminus E_m$ by the formula

$$f(x) = \lim_{j \to \infty} f_{n_j}(x) = f_{n_1}(x) + \lim_{k \to \infty} \sum\nolimits_{j=1}^{k} \left(f_{n_{j+1}}(x) - f_{n_j}(x)\right).$$

According to the Weierstrass M-test (see [43], Theorem 6.8.6, p. 172), the sequence $(f_{n_j})_j$ converges uniformly to f in $\Omega \setminus E_m$. As $(E_m)_{m \geq 1}$ is non-increasing, the function f is well-defined on $\bigcup_{m=1}^{\infty} (\Omega \setminus E_m)$. Put $f(x) = 0$ for $x \in \complement \left(\bigcup_{m=1}^{\infty} (\Omega \setminus E_m)\right) = \bigcap_{m=1}^{\infty} E_m$. Since, by the countable subadditivity of μ,

$$\mu(E_m) \leq \sum\nolimits_{j \geq m} \mu(F_j) \leq \frac{1}{2^{m-1}} \to 0 \tag{3.5}$$

as $m \to \infty$, we infer that $f_{n_j} \to f$ q.e. According to Theorem 3.3.1, this implies the measurability of f.

Taking into consideration that $\left|f_{n_j} - f_{n_i}\right|^p \to \left|f_{n_j} - f\right|^p$ q.e. as $i \to \infty$, it follows from Theorem 3.4.1 (iv) that

$$(\mathrm{C})\int_\Omega \left|f_{n_j} - f\right|^p \mathrm{d}\mu \leq \liminf_{i\to\infty} \left((\mathrm{C})\int_\Omega \left|f_{n_j} - f_{n_i}\right|^p \mathrm{d}\mu\right) \leq \left(\frac{1}{2^j}\right)^{p+1}$$

for every $j \in \mathbb{N}$. Therefore, $f_{n_j} - f \to 0$ in $L^p(\mu)$ and for each j the function $f_{n_j} \dot{-} f$ belongs to $L^p(\mu)$. Since $L^p(\mu)$ is a vector space, it follows that $f = \left(f - f_{n_j}\right) + f_{n_j}$ also belongs to $L^p(\mu)$. Then $f_{n_j} \to f$ in $L^p(\mu)$ and this fact combined with the property that $(f_n)_n$ is a Cauchy sequence in $L^p(\mu)$ implies the convergence of the entire sequence $(f_n)_n$ to f in $L^p(\mu)$. ■

Corollary 3.5.5 *Under the hypotheses of Theorem 3.5.4, the Banach spaces $L^p(\mu)$ $(1 \leq p < \infty)$ can be defined as the completion of the vector lattice $\mathcal{M}_b(\Omega, \Sigma)$ with respect to the norm $\|\cdot\|_{L^p(\mu)}$. This implies that all these spaces are actually Banach lattices.*

Taking into consideration the result of Lemma 2.5.5 about the uniform approximation of bounded measurable functions by simple functions, one can reformulate the last corollary as follows:

Corollary 3.5.6 *Under the hypotheses of Theorem* 3.5.4, *the set of simple functions is dense in each of the spaces $L^p(\mu)$ with $1 \leq p < \infty$.*

A careful inspection of the proof of Theorem 3.5.4 yields additional information about the convergence in a space $L^p(\mu)$. To detail, we shall need a definition.

Definition 3.5.7 A sequence of measurable functions $f_n : \Omega \to \overline{\mathbb{R}}$ converges quasi-uniformly to a function $f : \Omega \to \overline{\mathbb{R}}$ (equivalently, $f_n \to f$ q.u.) if for every $\varepsilon > 0$, there exists a set $E \in \Sigma$ such that $\mu(E) < \varepsilon$ and the functions f_n are finite and converge uniformly to f on $\Omega \setminus E$.

Using the notion of quasi-uniform convergence, one can check easily that the subsequence $(f_{n_j})_j$ constructed in the proof of Theorem 3.5.4 converges quasi-uniformly to f. In other words, the following is true:

Theorem 3.5.8 *Under the hypotheses of Theorem* 3.5.4, *if $f_n \to f$ in a space $L^p(\mu)$ with $p \in [1, \infty)$, then there is a subsequence of $(f_n)_n$ which converges to f quasi-uniformly.*

A classical result in measure theory known as the *Egorov Theorem* asserts that every almost everywhere pointwise convergent sequence of measurable functions is also quasi-uniform convergent on sets of finite measure. See Bogachev [25], Theorem 2.2.1, p. 110.

However, this is not true for capacities in general, a counterexample being made available by Ponce and Spector. See [151], Example 2.2.

Let us mention yet another consequence of the convergence in a space $L^p(\mu)$.

Theorem 3.5.9 *If $f_n \to f$ in a space $L^p(\mu)$ with $p \in [1, \infty)$, then $f_n \to f$ in capacity, i.e.,*

$$\mu\{x : |f_n(x) - f(x)| > \varepsilon\} \to 0 \quad \text{as } n \to \infty,$$

for every $\varepsilon > 0$.

Proof This is a consequence of the fact that

$$\mu\{x : |f_n(x) - f(x)|^p > \varepsilon\} \leq \frac{1}{\varepsilon} \text{(C)} \int_{\Omega} |f - f_n|^p \, \mathrm{d}\mu$$

for every $\varepsilon > 0$. ■

The scale of L^p spaces can be completed by adding the capacitary analog of the Lebesgue space L^∞.

$\mathcal{L}^\infty(\mu)$ is the space of all measurable functions such that

$$\|f\|_{L^\infty(\mu)} = \inf\{M > 0 : |f| \leq M \text{ q.e.}\} < \infty.$$

Since μ is strongly subadditive, $\mathcal{L}^\infty(\mu)$ is a vector space, and $\|\cdot\|_{L^\infty(\mu)}$ is a seminorm. The space $L^\infty(\mu)$ is defined as the space of equivalence classes $[f]$ of functions f in $\mathcal{L}^\infty(\mu)$, relative to the equivalence relation

$$f \sim g \text{ if and only if } f = g \text{ q.e.}$$

We will refer to $L^\infty(\mu)$ as the space of all *essentially bounded functions*.

Convergence in $L^\infty(\mu)$ is the same as the q.u. convergence.

Proposition 3.5.10 *If $(f_n)_n$ converges to 0 in $L^\infty(\mu)$, then there is a subset $E \subset \Omega$ of capacity zero such that $(f_n)_n$ converges uniformly to 0 on $\Omega \setminus E$.*

Proof For each n, there is a subset $E_n \subset \Omega$ such that $\mu(E_n) = 0$ and $|f_n(x)| \leq \|f_n\|_{L^\infty(\mu)}$ for all $x \in \Omega \setminus E_n$. Since μ is countably subadditive, it follows that we can take $E = \bigcup_n E_n$. The proof ends by noticing that

$$\sup_{x \in \Omega \setminus E} |f_n(x)| \leq \sup_{x \in \Omega \setminus E_n} |f_n(x)| \leq \|f_n\|_{L^\infty(\mu)} \to 0$$

as $n \to \infty$. ■

Theorem 3.5.11 *The space $L^\infty(\mu)$ is complete.*

Proof Let $(f_n)_n$ be a Cauchy sequence in $L^\infty(\mu)$. Then, repeating the argument of Proposition 3.5.10, one deduces the existence of a set E of capacity zero such that $(f_n)_n$ is uniformly Cauchy on $\Omega \setminus E$. As a consequence, $(f_n)_n$ is uniformly convergent on $\Omega \setminus E$ to a function f. If we define $f(x) = 0$ on E,, one obtains a bounded measurable function to which $(f_n)_n$ converges in $L^\infty(\mu)$. Indeed,

$$\|f_n - f\|_{L^\infty(\mu)} \leq \sup_{x\in\Omega\setminus E} |f_n(x) - f(x)| \to 0$$

as $n \to \infty$. ■

Several other important results are mentioned below. Notice that according to our assumptions made at the beginning of this section, μ is a finite capacity.

Theorem 3.5.12 (The Scale of L^p Spaces) *If $1 \leq p \leq q \leq \infty$,, then*

$$L^\infty(\mu) \subset \cdots \subset L^q(\mu) \subset L^p(\mu) \subset \cdots \subset L^1(\mu),$$

as linear spaces. Moreover, these inclusions are dense (see Theorem 3.5.1) and continuous since

$$\|f\|_{L^p(\mu)} \leq \mu(\Omega)^{1/p} \|f\|_{L^\infty(\mu)} \quad \text{if } f \in L^\infty(\mu) \text{ and } p \geq 1$$

and

$$\|f\|_{L^p(\mu)} \leq \mu(\Omega)^{1/p-1/q} \|f\|_{L^q(\mu)} \quad \text{if } f \in L^q(\mu)\ 1 < p < q < \infty,$$

according to Hölder's inequality (cf. Theorem 3.1.6).

In connection to the scale of L^p spaces, let us mention the following *interpolation property* of L^p norms.

Let $1 \leq p_0 < p_1 \leq \infty$ and $f \in L^{p_0}(\mu) \cap L^{p_1}(\mu)$. Then $f \in L^p(\mu)$ for all $p \in [p_0, p_1]$ and in addition

$$\|f\|_{L^{p_\theta}(\mu)} \leq \|f\|_{L^{p_0}(\mu)}^{1-\theta} \|f\|_{L^{p_1}(\mu)}^{\theta} \tag{3.6}$$

for all $\theta \in [0, 1]$, where the exponent p_θ is defined by $1/p_\theta = (1-\theta)/p_0 + \theta/p_1$. If $p_1 < \infty$, this follows from Hölder's inequality. Indeed,

$$\begin{aligned}\|f\|_{L^{p_\theta}(\mu)}^{p_\theta} &= (\mathrm{C})\int_\Omega |f|^{(1-\theta)p_\theta}\,|f|^{\theta p_\theta}\,\mathrm{d}\mu \\ &\le \left\| |f|^{(1-\theta)p_\theta} \right\|_{L^{p_0/[(1-\theta)p_\theta]}(\mu)} \left\| |f|^{\theta p_\theta} \right\|_{L^{p_1/(\theta p_\theta)}(\mu)} \\ &= \|f\|_{L^{p_0}(\mu)}^{(1-\theta)p_\theta} \, \|f\|_{L^{p_1}(\mu)}^{\theta p_\theta}\,.\end{aligned}$$

The case $p_1 = \infty$ is immediate, and we omit the details.

Theorem 3.5.13 *The space $L^\infty(\mu)$ is a commutative algebra and the product of any function $f \in L^p(\mu)$ with a function $g \in L^\infty(\mu)$ belongs to $L^p(\mu)$. In addition,*

$$\left|(\mathrm{C})\int_\Omega fg\,\mathrm{d}\mu\right| \le \mu(\Omega)^{1-1/p}\,\|f\|_{L^p(\mu)}\,\|g\|_{L^\infty(\mu)}\,.$$

Remark 3.5.14 As was mentioned in Sect. 2.4, the Choquet integrals of real-valued functions make sense only if they are computed against sets of finite capacity. This restriction does not apply in the case of positive functions. Taking into account the above discussion, one can easily check that the positive cones

$$L_+^p(\mu) = \left\{ f : \Omega \to \mathbb{R}_+ \cup \{\infty\} : f \text{ measurable and } (\mathrm{C})\int_\Omega f\,\mathrm{d}\mu < \infty \right\}$$

make sense for all lower continuous, strongly subadditive and complete capacities $\mu : \Sigma \to \mathbb{R}_+ \cup \{\infty\}$ (defined on a σ-algebra Σ of subsets of a total set Ω). These cones are complete and play (with obvious modifications) all nice approximation properties of L^p spaces as well as the interpolation inequality *(3.6)*. However, the Theorem *3.5.12*, concerning the scale of L^p spaces, does not work in this context (actually it fails even in the case of Lebesgue measure on $\mathbb{R}$).

3.6 Approximation by Continuous Functions with Compact Support

In the preceding section, the capacitary Lebesgue spaces $L^P(\mu)$ were defined in relation to measurable functions. When the state space Ω is a locally compact Hausdorff space and the σ-algebra under attention includes the Borel σ-algebra $\mathcal{B}(\Omega)$, it is natural to relate the discussion on the spaces $L^P(\mu)$ to continuity. The clue is offered by Luzin's Theorem about the continuity of a measurable functions on nearly all its domain.

Theorem 3.6.1 (Luzin's Theorem) *Given a Lebesgue measurable set $E \subset \mathbb{R}^N$ and a function $f : E \to \overline{\mathbb{R}}$ finite almost everywhere, the following statements are equivalent.*

(i) *f is measurable;*
(ii) *For each $\varepsilon > 0$, there exists an open set $U \subset E$ with $m(U) < \varepsilon$ such that $f|_{(E \setminus U)}$ is continuous in the relative topology that is,*

$$x_n, x \in E \setminus U \ (n \in \mathbb{N}) \text{ and } x_n \to x \Rightarrow f(x_n) \to f(x).$$

Interestingly, Luzin's Theorem admits an analog in the setting of complete capacity spaces (Ω, Σ, μ), where Ω is a locally compact Hausdorff space that can be represented as a countable union of compact sets and Σ is a σ-algebra that includes the Borel σ-algebra $\mathcal{B}(\Omega)$ (as well as all subsets of sets having capacity zero). We will need the following equivalent of measurability, suggested by Luzin's Theorem.

Definition 3.6.2 A function $f : \Omega \to \overline{\mathbb{R}}$ is quasi-continuous if for every $\varepsilon > 0$, there exists an open set $U \subset \Omega$ such that $\mu(U) < \varepsilon$ and $f|_{\Omega \setminus U}$ is finite and continuous *(*in the relative topology*)*.

Based on this definition, we can formulate the following extension of Theorem 3.6.1:

Theorem 3.6.3 (The Capacitary Variant of *Luzin's Theorem*) *Suppose that Ω is a locally compact Hausdorff space that can be represented as a countable union of compact sets, Σ is a σ-algebra that includes the Borel σ-algebra $\mathcal{B}(\Omega)$ and $\mu : \Sigma \to \mathbb{R}_+ \cup \{\infty\}$ is a complete capacity that verifies the following regularity conditions:*

(i) $\mu(K) < \infty$ *for each compact set K;*
(ii) $\mu(A) = \inf\{\mu(U) : A \subset U$ *and U is open*$\}$ *for each Borel set A;*
(iii) $\mu(A) = \sup\{\mu(K) : K \subset A$ *and K is compact*$\}$ *for each Borel subset A of finite capacity.*

Then, a function $f : \Omega \to \overline{\mathbb{R}}$ q.e. finite is measurable if and only if it is quasi-continuous.

For example, the hypotheses of Theorem 3.6.3 are fulfilled by every distortion of the Lebesgue measure on $\mathbb{R}^N$ (see Example 2.1.7) as well as by the capacities mentioned in Sect. 2.2.

Proof Suppose first that $f : \Omega \to \overline{\mathbb{R}}$ is a q.e. finite measurable function and $\mu(\Omega) < \infty$. Modifing f at all points with infinite values we may assume that f is real-valued. See Corollary 3.3.2.

Let $\varepsilon > 0$ arbitrarily fixed, and let $(V_n)_n$ be an enumeration of the open intervals in $\mathbb{R}$ with rational endpoints. According to the regularity properties (i) and (ii), there exists compact sets K_n and open sets U_n such that

$$K_n \subset f^{-1}(V_n) \subset U_n \text{ and } \mu(U_n \setminus K_n) < \varepsilon/2^n.$$

Put $C = \Omega \setminus \left(\bigcup_{n=1}^{\infty} (U_n \setminus K_n)\right)$. Then C is closed and $U = \Omega \setminus C$ is an open set such that

$$\mu(U) \leq \sum_{n=1}^{\infty} \mu_n \left(U_n \setminus K_n\right) < \varepsilon,$$

due to the countable subadditivity of μ. The continuity of the restriction of f to C follows from the fact that

$$(f|_C)^{-1}(V_n) = C \cap f^{-1}(V_n) = C \cap U_n,$$

due to the definition of C.

Consider now the case when $\mu(\Omega) = \infty$ and choose an increasing sequence $(K_n)_n$ of compact sets such that $\Omega = \bigcup_{n=1}^{\infty} K_n$. According to the discussion above, for every $\varepsilon > 0$ and $n \in \mathbb{N}$, there exist open sets W_n such that $\mu(W_n) < \varepsilon/2^n$ and $f|_{K_n \setminus W_n}$ is continuous. Then $W = \bigcup_{n=1}^{\infty} W_n$ is an open set and $\mu(W) < \varepsilon$, while f is continuous on $\bigcup_{n=1}^{\infty} (K_n \setminus W_n) = \Omega \setminus W$. In other words, f is quasi-continuous.

Conversely, suppose that f is quasi-continuous. Then for each $n \in \mathbb{N}$, there exists an open set $U_n \subset \Omega$ such that $\mu(U_n) < 1/n$ and $f|_{\Omega \setminus U_n}$ is continuous. The set $\bigcap_{n=1}^{\infty} U_n$ is a Borel set and thus measurable. Moreover $\mu\left(\bigcap_{n=1}^{\infty} U_n\right) = 0$.

The set Ω decomposes into a closed set

$$C = \bigcup_{n=1}^{\infty} (\Omega \setminus U_n) = \Omega \setminus \left(\bigcap_{n=1}^{\infty} U_n\right)$$

and a set $Z = \bigcap_n U_n$ of capacity zero. Then, for every $\alpha \in \mathbb{R}$, we have

$$\begin{aligned}\{x \in \Omega : f(x) > \alpha\} &= \{x \in C : f(x) > \alpha\} \cup \{x \in Z : f(x) > \alpha\} \\ &= \bigcup_{n=1}^{\infty} \{x \in \Omega \setminus U_n : f(x) > \alpha\} \cup \{x \in Z : f(x) > \alpha\}.\end{aligned}$$

Since $f|_{\Omega \setminus U_n}$ is continuous, the set $\{x \in \Omega \setminus U_n : f(x) > \alpha\}$ is the intersection of an open subset of Ω with the closed set $\Omega \setminus U_n$ and thus is measurable. The set $\{x \in Z : f(x) > \alpha\}$ is a subset of a set of capacity zero, and it is also measurable since μ is a complete capacity. In conclusion, all the sets $\{x \in \Omega : f(x) > \alpha\}$ are measurable, which shows that f is measurable. ■

Corollary 3.6.4 *If μ is a complete capacity, which verifies the hypotheses of Theorem 3.6.3, then the spaces $L^p(\mu)$ $(1 \leq p < \infty)$ can be characterized as the spaces of (classes modulo equality q.e.) of quasi-continuous functions f such that* $(C)\int_\Omega |f|^p \, d\mu < \infty$.

The approximation of integrable functions by continuous functions makes the objective of the following proposition.

Proposition 3.6.5 *Suppose that* (Ω, Σ, μ) *is a capacity space, where* Ω *is a metric space,* Σ *is a* σ*-algebra that includes the Borel* σ*-algebra* $\mathcal{B}(\Omega)$*, and* μ *is a complete and subadditive capacity defined on* Σ*. If* $f : \Omega \to \overline{\mathbb{R}}$ *is quasi-continuous and* (C) $\int_\Omega |f|^p \,\mathrm{d}\mu < \infty$*, then there exists a sequence* $(f_n)_n$ *of bounded continuous functions in* Ω *such that*

$$\lim_{n\to\infty} \text{(C)} \int_\Omega |f - f_n|^p \,\mathrm{d}\mu = 0.$$

Proof Repeating the argument of Theorem 3.5.1, for every $\varepsilon > 0$,, there is a quasi-continuous and bounded function g, such that $\|f - g\|_{L^p(\mu)} < \varepsilon$ and $|g| \leq M$ for some $M > 0$.

Since g is quasi-continuous, there exists an open set U such that $\mu(U) < \varepsilon^p/(2M)^p$ and $g|_{\Omega\setminus U}$ is continuous. By the Tietze extension theorem (see the Comments at the end of this chapter), one can find a continuous function $h : \Omega \to \mathbb{R}$ with $|h| \leq M$ and $h|_{\Omega\setminus U} = g$. Then

$$\begin{aligned}
\|g - h\|^p_{L^p(\mu)} &= \text{(C)} \int_\Omega |g - h|^p \,\mathrm{d}\mu = \int_0^\infty \mu\left(\{x \in \Omega : |g - h|^p > t\}\right) \mathrm{d}t \\
&= \int_0^\infty \mu\left(\{x \in U : |g - h|^p > t\}\right) \mathrm{d}t \\
&= \int_0^{(2M)^p} \mu\left(\{x \in U : |g - h|^p > t\}\right) \mathrm{d}t \\
&\leq (2M)^p \mu(U) < \varepsilon^p.
\end{aligned}$$

Finally, taking into account that μ is subadditive, we infer from Remark 3.1.11 that

$$\|f - h\|_{L^p(\mu)} \leq 2\left(\|f - g\|_{L^p(\mu)} + \|g - h\|_{L^p(\mu)}\right) < 2\varepsilon,$$

which ends the proof. ■

An important problem is when $C_c(\mathbb{R}^N)$ (the space of continuous functions $f : \mathbb{R}^N \to \mathbb{R}$ with compact support) is dense into each of the spaces $L^p(\mu)$ $(1 \leq p < \infty)$ attached to a capacity $\mu : \mathcal{P}(\mathbb{R}^N) \to [0, \infty]$.

To answer this problem, notice first that the Choquet integral of every function $f \in C_c(\mathbb{R}^N)$ is finite provided that μ is *locally finite* in the sense that

$$\mu(U) < \infty \text{ for every bounded open set } U \subset \mathbb{R}^N.$$

As noticed Ponce and Spector [151], the density of the space $C_c(\mathbb{R}^N)$ is possible only for those capacities that verify in addition the following vanishing property at infinity.

Definition 3.6.6 A capacity $\mu : \mathcal{P}(\mathbb{R}^N) \to [0, \infty]$ is evanescent if for every open subset $U \subset \Omega$ with finite capacity and every closed subset $C \subset U$, we have

$$\lim_{r\to\infty} \mu(C \setminus B_r(0)) = 0.$$

The distorted Lebesgue measures (as well as the Riesz and Besov capacities) are examples of evanescent capacities.

Theorem 3.6.7 *A subadditive capacity $\mu : \mathcal{P}(\mathbb{R}^N) \to [0, \infty]$ is evanescent if and only if for every quasi-continuous and pth μ-integrable function $f : \mathbb{R}^N \to \overline{\mathbb{R}}$, there exists a sequence of functions $f_n \in C_c(\Omega)$ such that*

$$\lim_{n\to\infty} (\mathrm{C}) \int_{\Omega} |f - f_n|^p \, \mathrm{d}\mu = 0.$$

For details, see Ponce and Spector [151], Proposition 3.3 and Proposition 3.4. It is important to mention that their argument makes use of a very special property of $\mathbb{R}^N$, the compactness of every closed and bounded set.

Corollary 3.6.8 *Suppose that $\mu : \mathcal{P}(\mathbb{R}^N) \to [0, \infty]$ is a strongly subadditive, countably subadditive, evanescent, and locally finite capacity. Then, $L^p(\mu)$ is the completion of the space (of classes of functions in) $C_c(\mathbb{R}^N)$ with respect to the $L^p(\mu)$ norm.*

3.7 Comments

Chebyshev's probabilistic inequality was published by Chebyshev [39] in a paper from 1867, where he proved the weak law of large numbers for arithmetic means.

The extension of Jensen's inequality to the framework of Choquet integral (as stated in Theorem 3.1.10) can be traced back to an old result due to Jessen [101], who considered the case of positive linear functionals A defined on an ordered vector space of functions, normalized by the condition $A(1) = 1$.

The extension of Hölder's inequality to the framework of Choquet integral was treated by numerous authors, see, for example, Agahi [4], Cerdà et al. [34], Mesiar, Li and Pap [128], and Wang [182].

Luzin's theorem, which asserts the almost everywhere continuity of Lebesgue measurable functions, is omnipresent in all courses dedicated to measure theory and integration. Ferguson and Wu [64] succeeded "a one-sentence inverse image proof" of it. Our extension of this result to the framework of capacities (stated as Theorem 3.6.3) is based on their ideas. As noticed Wiesel [186], in the context of subadditive capacities μ defined on the Borel sets of a compact metric space, Luzin's theorem is equivalent to the fact that

$$\lim_{n\to\infty} \mu(O_n) = \mu\left(\bigcap\nolimits_{n=1}^{\infty} O_n\right)$$

for every decreasing sequence $(O_n)_n$ of open sets.

The Tietze Extension Theorem is a basic result in Function Theory and Topology. Tietze settled the case of metric spaces as follows: Suppose M is a metric space and S is a closed subset of M. Suppose $f \in C(S, \mathbb{R})$. Then there exists $\tilde{f} \in C(M, \mathbb{R})$ such that $\tilde{f}|_S = f$ and $\inf \tilde{f}(M) = \inf f(S)$ and $\sup \tilde{f}(M) = \sup f(S)$. A simple proof was made available by O'Searcoíd [147], Theorem 10.9.4, p. 178. A powerful variant of Tietze's Theorem to the context of locally compact Hausdorff spaces made the objective of Theorem 7.40, p. 99, in the monograph of Hewitt and Stromberg [98].

Several constructions of L^p spaces within the Choquet integral (with varying degrees of generality) can be found in the works of Adams [2], Adams and Hedberg [3], Cerreia-Vioglio et al. [35], Denneberg [51], Harjulehto and Hurri-Syrjänen [97], Ooi [144, 145], Ponce and Spector [151], and Zhou [195], to cite just a few.

Supplementary details concerning the construction of the Lebesgue capacitary spaces in the case of c-subadditive capacities are made available by the paper of Cerdà, Martín and Silvestre [34].

4 Weakly Nonlinear Operators Acting on Ordered Banach Spaces

As was noticed in the preceding chapter, the Choquet integral associates to every capacity space $(\Omega, \mathcal{A}, \mu)$ an integral functional,

$$I(f) = (\mathrm{C}) \int_{\Omega} f \, \mathrm{d}\mu, \quad f \in \mathcal{M}_b(\Omega, \mathcal{A}),$$

which we referred to as the Choquet functional. This functional, though nonlinear, displays a number of important properties like comonotonic additivity, increasing monotonicity and positive homogeneity, which actually characterizes it among the functionals $I : \mathcal{M}_b(\Omega, \mathcal{A}) \to \mathbb{R}$. Moreover, if $\mathcal{A}$ is a σ-algebra, then $\mathcal{M}_b(\Omega, \mathcal{A})$ is a linear space and I is sublinear if (and only if) μ is strongly subadditive.

Choquet functionals provide a basic example for a whole class of operators that attracted our attention when extending Korovkin's approximation theorems to a nonlinear framework. See [75, 76] and [77]. These operators, which combine the increasing monotonicity, sublinearity, and a weaker form of translation invariance, deserve to be studied on their own. A comprehensive introduction to their study is the goal of this chapter.

4.1 Generalities on the Weakly Nonlinear Operators

Throughout this section, we consider operators defined on ordered Banach spaces E that admits an abstract unity $u > 0$ and take values into an ordered Banach space F. The term *abstract unity* refers to a point specified according to a given context. For example, we make the convention to chose $u = 1$ whenever E embeds into a function space $\mathcal{F}(\Omega)$ and E contains the unity 1. So is the case of the spaces $\mathbb{R}^N$, c, ℓ^∞, $C(K)$, $C_b(\Omega)$, $\mathcal{UC}_b(\Omega)$, $\mathcal{F}_b(\Omega)$, $\mathcal{M}_b(\Omega, \Sigma)$ and of the Lebesgue spaces $L^p(\mu)$ with $1 \leq p \leq \infty$, associated with

S. G. Gal, C. P. Niculescu, *Choquet Integral and Monotone Sublinear Operators*, Frontiers in Mathematics, https://doi.org/10.1007/978-3-031-71701-7_4

a finite measure μ. In the case of matrix spaces like $\mathrm{Sym}(N, \mathbb{R})$ (or of the operator spaces like $B_{sa}(H)$), the usual choice for u is the identity I. Any other situations will be explicitly mentioned.

Definition 4.1.1 An operator $T : E \to F$ is said to be *weakly nonlinear* if it satisfies the following two conditions:

(SL) *(Sublinearity)* T is subadditive and positively homogeneous, i.e.,

$$T(f+g) \leq T(f) + T(g) \quad \text{and} \quad T(\alpha f) = \alpha T(f)$$

for all f, g in E and $\alpha \geq 0$;

(TR) *(Translatability)* $T(f + \alpha \cdot u) = T(f) + \alpha T(u)$ for all functions $f \in E$ and all numbers $\alpha \geq 0$.

When $E = F$ and T is *unital* (i.e., $T(u) = u$), the condition of translatability takes the form

$$T(f + \alpha u) = T(f) + \alpha u,$$

for all $f \in E$ and $\alpha \geq 0$.

A stronger condition than translatability is the following one:

(INV) (*Translation invariance*) $T(f+\alpha u) = T(f)+\alpha T(u)$ *for all functions* $f \in E$ *and all numbers* $\alpha \in \mathbb{R}$.

The last condition occurs naturally (for $u = 1$) in the context of Choquet integral, being a consequence of the property of *comonotonic additivity*, i.e.,

(CA) $T(f+g) = T(f) + T(g)$ *whenever the functions* $f, g \in E$ *are comonotone.*

In this book, we are especially interested in those weakly nonlinear operators which preserve the ordering, that is, which verify the following condition:

(M) (*Increasing monotonicity*) $T(f) \leq T(g)$ whenever $f \leq g$.

The last definition deserves an explanation. Every sublinear operator is convex, and in Convex Analysis, the notion of monotone operator has a different meaning, related to the behavior of its subgradients. See [137]. That's why we will use the term of monotone increasing operator and not of monotone operator.

Definition 4.1.2 The comonotonic additive and monotone increasing operators will be referred to as Choquet-type operators (respectively Choquet-type functionals, if the codomain is $\mathbb{R}$).

Notice that every Choquet-type operator is positively homogeneous, but not necessarily sublinear.

As it is well known, every sublinear operator $T : E \to F$ verifies the conditions $T(0) = 0$ and

$$-T(-f) \leq T(f) \quad \text{for all } f \in E.$$

As a consequence, every sublinear and monotone increasing operator T maps positive elements into positive elements, that is,

$$Tf \geq 0 \quad \text{for all } f \geq 0. \tag{4.1}$$

The property (4.1) is equivalent to the increasing monotonicity in the case of linear operators.

Remark 4.1.3 If T is a weakly nonlinear and monotone increasing operator, then

$$T(\alpha \cdot u) = \alpha \cdot T(u) \quad \text{for all } \alpha \in \mathbb{R}.$$

Indeed, for $\alpha \geq 0$ the property follows from positive homogeneity. Suppose now that $\alpha < 0$. Since $T(0) = 0$ and $-\alpha > 0$, by translatability, it follows that $0 = T(0) = T(\alpha \cdot u + (-\alpha) \cdot u) = T(\alpha \cdot u) + (-\alpha)T(u)$, which implies $T(\alpha u) = \alpha T(u)$.

Crandall and Tartar [45] have noticed that if an operator T, from $L^\infty(\mu)$ into itself, is translatable and unital, then it is monotone increasing if and only if it is Lipschitz with Lipschitz constant at most 1. In particular, this remark applies to any operator acting on ℓ^∞ which verifies the property (INV).

Remark 4.1.4 Important to notice is the fact that all linear, positive, and unital operators illustrate the class of weakly nonlinear and monotone increasing operators (and also the class of Choquet-type operators).

Real Analysis, Ergodic Theory, Harmonic Analysis, and Probability Theory offer numerous examples of such nonlinear operators. Few of them are presented in what follows.

Example 4.1.5 (The Limit Superior Functional) According to Example 2.8.5, the functional

$$I : \ell^\infty \to \mathbb{R}, \quad I((x_n)_n) = \limsup_{n\to\infty} x_n,$$

constitutes an example of weakly nonlinear and unital Choquet functional. The same is true for the functional

$$J : \ell^\infty \to \mathbb{R}, \quad J((x_n)_n) = \sup_{n\in\mathbb{N}} x_n.$$

Example 4.1.6 Suppose that Ω is an open subset of $\mathbb{R}^N$ and consider the *signed Hardy–Littlewood maximal operator* M_Ω, acting from the Lebesgue space $L^p(\Omega)$ ($p \in (1,\infty)$) into itself via the formula

$$(M_\Omega f)(x) = \sup \frac{1}{\operatorname{vol}(B_r(x))} \int_{B_r(x)} f(y)\mathrm{d}y, \tag{4.2}$$

where the supremum is taken over all radii $r > 0$ with $0 < r < \operatorname{dist}(x, \partial\Omega)$. Clearly, M_Ω is weakly nonlinear and monotone increasing. Besides, M_Ω is unital if Ω is bounded.

Notice that the Hardy–Littlewood maximal operator $M : L^p(\mathbb{R}^N) \to L^p(\mathbb{R}^N)$ ($p > 1$), defined by the formula

$$Mf(x) = \sup_{r>0} \frac{1}{\operatorname{vol}(B_r(x))} \int_{B_r(x)} |f(y)|\,\mathrm{d}y,$$

is neither monotone nor translatable.

Example 4.1.7 The operator $S : L^p(\mu) \to L^p(\mu)$ ($p \in (1,\infty)$) associated with a Borel probability measure μ on $(0, 1)$ via the formula

$$(Sf)(t) = \sup_{\mu(A)\le t} \frac{1}{\mu(A)} \int_A f(s)\mathrm{d}\mu,$$

is another example of weakly nonlinear, monotone increasing, and unital operator.

Example 4.1.8 Suppose now that E is a vector lattice of functions that contains the unity and $U : E \to E$ is a weakly nonlinear, monotone increasing, and unital operator. Then each of the *Yosida-Kakutani operators*,

$$YK_n : E \to E, \quad YK_n(f) = \sup\left\{ f, \frac{1}{2}(f + Uf), \ldots, \frac{1}{n}\sum_{k=0}^{n-1} U^k f \right\} \quad (n \in \mathbb{N})$$

is also weakly nonlinear, monotone increasing, and unital. In the case when $E = L^p(\mu)$ (with $1 \leq p < \infty$ and μ is a probability measure) and U verifies in addition the condition

$$\int |U(f)|^p \, d\mu \leq C \int |f|^p \, d\mu \quad \text{for all } f,$$

then the operators YK_n appear as truncations of the operator

$$YK : E \to E, \quad YK(f) = \sup \left\{ f, \frac{1}{2}(f + Uf), \ldots, \frac{1}{n}\sum_{k=0}^{n-1} U^k f, \ldots \right\},$$

which is also weakly nonlinear, monotone increasing, and unital. The operators YK_n and YK are related to the maximal ergodic theorem of Yosida and Kakutani [191].

The set $\mathcal{WN}(C_b(X), C_b(X))$ of all weakly nonlinear operators T from $C_b(X)$ into itself is a closed convex cone in the Banach space $\mathrm{Lip}_0\,(C_b(X), C_b(X))$ of all Lipschitz maps that vanishes at the origin of $C_b(X)$; the natural norm on this space is

$$\|T\|_{\mathrm{Lip}} = \sup \left\{ \frac{\|T(f) - T(g)\|}{\|f - g\|} : f, g \in C_b(X),\ f \neq g \right\}.$$

In turn, $\mathcal{WN}(C_b(X), C_b(X))$ contains the positive cone of the ordered Banach space $L(C_b(X), C_b(X))$.

A general procedure to generate new weakly nonlinear operators from old ones is as follows:

Lemma 4.1.9 (i) *If $S, T \in \mathcal{WN}(C_b(X), C_b(X))$ and $S(1) = T(1)$, then the operator $S \vee T$ defined by the formula*

$$(S \vee T)(f) = \sup\{S(f), T(f)\} \quad \text{for } f \in C_b(X),$$

also belongs to $\mathcal{WN}(C(X), C(X))$.

(ii) *If $S, T \in \mathcal{WN}(C_b(X), C_b(X))$ and T is unital, then*

$$ST \in \mathcal{WN}(C_b(X), C_b(X)).$$

Proof (i) Indeed, the fact that the pointwise sup of two sublinear operators is also a sublinear operator is obvious. In addition,

$$\begin{aligned}(S \vee T)(f + \alpha 1) &= \sup\{S(f + \alpha 1), T(f + \alpha 1)\} \\ &= \sup\{S(f) + \alpha S(1), T(f) + \alpha T(1)\}\end{aligned}$$

$$= \sup\{S(f), T(f)\} + \sup\{\alpha S(1), \alpha T(1)\}$$
$$= \sup\{S(f), T(f)\} + \alpha\,(S \vee T)\,(1)$$

for all $f \in C(X)$ and $\alpha \geq 0$. The assertion (ii) is straightforward. ■

According to Lemma 4.1.9 (i), the following operators, from c into itself, are unital and weakly nonlinear:

$$T_1\left((x_n)_n\right) = \left(\sup\{x_n, \lim_{k\to\infty} x_k\}\right)_n,$$

$$T_2\left((x_n)_n\right) = \left(\sup\left\{\frac{x_1 + \cdots + x_n}{n}, \lim_{k\to\infty} x_k\right\}\right)_n,$$

$$T_3\left((x_n)_n\right) = \left(\sup\left\{x_n, \frac{x_1 + 2x_2 + \cdots + 2^n x_n}{2^{n+1} - 1}\right\}\right)_n.$$

We end this section with a few remarks on the continuity of sublinear and monotone increasing operators.

Theorem 4.1.10 *Suppose that E and F are two ordered Banach spaces. Then every sublinear and monotone increasing operator $T : E \to F$ is Lipschitz continuous.*

Before entering the details, notice that the continuity at the origin of a sublinear operator $T : E \to F$ is equivalent to the existence of a constant $C \geq 0$ such that

$$\|T(f)\| \leq C\,\|f\| \quad \text{for all } f \in E;$$

to show this, repeat the standard argument used in the case of linear operators. The smallest such constant $C \geq 0$ will be called the *norm* of T and denoted by $\|T\|$. Thus

$$\|T\| = \inf\{\lambda > 0 : \|T\,(f)\| \leq \lambda\,\|f\| \text{ for all } f \in E\}$$
$$= \sup\{\|T(f)\| : f \in E,\ \|f\| \leq 1\}.$$

With this preparation, the assertion of Theorem 4.1.10 follows from the next two lemmas.

Lemma 4.1.11 *Every sublinear and monotone increasing operator $T : E \to F$ acting on ordered Banach spaces is continuous at the origin.*

Proof Without loss of generality, we may assume that E is a regularly ordered Banach space. See Lemma 1.2.1. If T were not continuous, then would exist a sequence $(f_n)_n$ of elements of E such that $\|f_n\| \leq 1/(n2^{n+1})$ and $\|T(f_n)\| \geq n$. According to the formula

$$\|f_n\| = \inf\{\|g_n\| : -g_n \leq f_n \leq g_n\},$$

one may choose elements g_n such that $-g_n \leq f_n \leq g_n$ and $\|g_n\| \leq 1/(n2^n)$ for every n. Since T is monotone increasing,

$$T(g_n) \geq T(f_n) \geq T(-g_n) \geq -T(g_n),$$

which implies that $\|T(g_n)\| \geq \|T(f_n)\| \geq n$. The series $\sum g_n$ being absolutely convergent is also convergent, with sum $g \geq 0$. Since T is monotone increasing, $T(g) \geq T(g_n)$, which implies $\|T(g)\| \geq \|T(g_n)\| \geq n$, impossible for n sufficiently large. In conclusion, the operator T must be continuous at the origin. ■

Lemma 4.1.12 *The continuity at the origin of a sublinear and monotone increasing operator* $T : E \to F$ *acting on ordered Banach spaces implies its Lipschitzianity.*

Proof According to Lemma 1.2.1, we may assume that F is a regularly ordered Banach space.

Let $f, g \in E$ and $y^* \in F_+^*$ be arbitrarily chosen. Since T is sublinear and y^* is monotone increasing, it follows that

$$y^*(T(f)) \leq y^*(T(g)) + y^*(T(f-g))$$

and

$$y^*(T(g)) \leq y^*(T(f)) + y^*(T(g-f)).$$

Then

$$\begin{aligned}
\left|y^*(T(f) - T(g))\right| &\leq \max\left\{\left|y^*(T(f-g))\right|, \left|y^*(T(g-f))\right|\right\} \\
&\leq \left\|y^*\right\| \max\{\|T(f-g)\|, \|T(g-f)\|\} \\
&\leq \left\|y^*\right\| \|T\| \|f-g\|.
\end{aligned}$$

According to the assertions (i) and (ii) of Lemma 1.2.2, there is a constant $C > 0$ such that every functional $z^* \in F^*$ admits a decomposition of the form $z^* = u^* - v^*$ with $u^*, v^* \in F^*$ and $\|u^*\|, \|v^*\| \leq C\|z^*\|$. Therefore,

$$\left|z^*(T(f) - T(g))\right| \leq \left|u^*(T(f) - T(g))\right| + \left|v^*(T(f) - T(g))\right|$$

$$\leq 2C \left\|z^*\right\| \|T\| \|f - g\|,$$

whence

$$\|T(f) - T(g)\| \leq 2C \|T\| \|f - g\|,$$

since z^* was arbitrarily chosen in F^*. ■

Lemma 4.1.13 *Suppose that E and F are two Banach lattices. If $T : E \to F$ is a continuous sublinear operator, then*

$$\|T(f) - T(g)\| \leq 2 \|T\| \|f - g\| \quad \textit{for all } f \in E.$$

Proof Indeed, for all $f, g \in E$, we have $T(f) \leq T(g) + T(f - g) \leq T(g) + |T(f - g)|$ and $T(g) \leq T(f) + |T(g - f)|$, whence

$$\begin{aligned}|T(f) - T(g)| &= \sup \{-(T(f) - T(g)), T(f) - T(g)\} \\ &\leq |T(f - g)| + |T(g - f)|.\end{aligned}$$

Therefore, $\|T(f) - T(g)\| \leq \|T(f - g)\| + \|T(g - f)\| \leq 2 \|T\| \|f - g\|$ and the proof is complete. ■

Remark 4.1.14 In the case of sublinear and monotone increasing operators $T : E \to F$ acting on Banach lattices, the result of Lemma 4.1.13 can be improved by noticing the inequality

$$|T(f) - T(g)| \leq T(|f - g|) \quad \text{for all } f, g \in E \tag{4.3}$$

and then, by applying norms to both sides, to conclude that the Lipschitz constant of T equals its norm,

$$\|T(f) - T(g)\| \leq \|T\| \|f - g\| \quad \text{for all } f, g \in E.$$

To prove 4.3, note that $f \leq g + |f - g|$ implies $T(f) - T(g) \leq T(|f - g|)$, due to the monotonicity and subadditivity of T. Interchanging the role of f, and g we infer that also $-(T(f) - T(g)) \leq T(|f - g|)$. Therefore,

$$|T(f) - T(g)| = \sup \{T(f) - T(g), -(T(f) - T(g))\} \leq T(|f - g|).$$

Remark 4.1.15 (The generality of the notions of sublinearity, translatability, and monotonicity) Clearly, sublinearity makes sense for all operators T defined on cones C (of real vector spaces E) and taking values in ordered vector spaces F. On the other hand, the property of translatability imposes the invariance of the domain of T relative to all mappings $T_\alpha : x \to x + \alpha u$ (with $\alpha \geq 0$). Thus, the notion of weakly nonlinear operator (and in particular that of monotone increasing weakly nonlinear operator) is perfectly motivated in the context of operators $T : C \to F$ defined on cones C of an ordered Banach space E that admits an abstract unity $u > 0$ such that $u \in C$.

The case of Choquet functionals (presented in Chap. 2) offers an illustration how these notions can be introduced when the domain is not closed to addition.

4.2 The Choquet–Bochner Integral

According to Schmeidler's representation Theorem 2.8.1, if I is a comonotonic additive and monotone increasing functional defined on $\mathcal{M}_b(\Omega, \mathcal{A})$, then the formula $\mu(A) = I(\chi_A)$ defines a capacity μ on $\mathcal{A}$ such that $\mu(\Omega) = I(1_\Omega)$ and

$$I(f) = (C)\int_\Omega f \, \mathrm{d}\mu \quad \text{for every } f \in \mathcal{M}_b(\Omega, \mathcal{A}).$$

Moreover (see Theorem 2.6.1), this functional is sublinear if and only if the capacity μ is strongly subadditive.

Now, if $T : \mathcal{M}_b(\Omega, \mathcal{A}) \to \mathbb{R}^N$ is a comonotonic additive and monotone increasing operator, then each of its components,

$$T_k : \mathcal{M}_b(\Omega, \mathcal{A}) \to \mathbb{R}, \quad T_k = \mathrm{pr}_k \circ T,$$

is an operator of the same nature which (by Schmeidler's Representation Theorem) can be represented by a capacity μ_k on $\mathcal{A}$ such that $\mu_k(\Omega) = T_k(1_\Omega)$ and

$$\left(\mathrm{pr}_k \circ T\right)(f) = (C)\int_\Omega f \, \mathrm{d}\mu_k \quad \text{for every } f \in \mathcal{M}_b(\Omega, \mathcal{A}).$$

Here $\mathrm{pr}_k(x_1, \ldots, x_k, \ldots, x_N) = x_k$ denotes the canonical projection to the kth component of a vector in $\mathbb{R}^N$. The capacities μ_k can be seen as the component of a vector capacity

$$\boldsymbol{\mu} : \mathcal{A} \to \mathbb{R}^N, \quad \boldsymbol{\mu}(A) = (\mu_1(A), \ldots, \mu_N(A))$$

such that $\boldsymbol{\mu}(\Omega) = T(1_\Omega)$ and

$$T(f) = \int_0^{+\infty} \boldsymbol{\mu}(\{x : f(x) \geq t\}) \, \mathrm{d}t + \int_{-\infty}^0 [\boldsymbol{\mu}(\{x : f(x) \geq t\}) - \boldsymbol{\mu}(\Omega)] \, \mathrm{d}t$$

for every $f \in \mathcal{M}_b(\Omega, \mathcal{A})$. Thus, by defining the integral $(C)\int_\Omega f\,\mathrm{d}\boldsymbol{\mu}$, of a function f with respect to the set function $\boldsymbol{\mu}$, as the sum of the two integrals appearing in the last formula, we actually preserve Schmeidler's Representation Theorem under the form

$$T(f) = (C)\int_\Omega f\,\mathrm{d}\boldsymbol{\mu} \quad \text{for every } f \in \mathcal{M}_b(\Omega, \mathcal{A}),$$

and $\boldsymbol{\mu}(\Omega) = T(1_\Omega)$.

The above discussion can be easily turned into a general construction of vector-valued comonotonic additive and monotone increasing operators, based on a fusion between the Bochner theory of integration of vector-valued functions and the Choquet integral.

As above, $\mathcal{A}$ denotes an algebra of subsets of an abstract set Ω and E is an ordered Banach space.

Definition 4.2.1 A set function $\boldsymbol{\mu} : \mathcal{A} \to E_+$ is called a *vector capacity* if it verifies the following two conditions:

(C1) $\boldsymbol{\mu}(\emptyset) = 0$; and
(C2) $\boldsymbol{\mu}(A) \le \boldsymbol{\mu}(B)$ for all $A, B \in \mathcal{A}$ such that $A \subset B$.

Notice that any vector capacity $\boldsymbol{\mu}$ is positive and takes values in the order interval $[0, \boldsymbol{\mu}(\Omega)]$.

An important class of vector capacities is that of *additive* (respectively σ*-additive*) *vector measures with positive values*, that is, of capacities $\boldsymbol{\mu} : \mathcal{A} \to E_+$ with the property that

$$\boldsymbol{\mu}\left(\bigcup\nolimits_{n\ge1} A_n\right) = \sum\nolimits_{n\ge1} \boldsymbol{\mu}(A_n),$$

for every finite (respectively countable) sequence $A_1, A_2, A_3, \ldots$ of disjoint sets belonging to $\mathcal{A}$, whose union also belongs to $\mathcal{A}$.

A vector capacity taking values in $\mathbb{R}_+^N$ is an additive/σ-additive measure if and only if its components have the respective property. An example involving the infinite dimensional spaces is provided by the formula

$$\boldsymbol{\mu}_p : \mathcal{B}([0,1]) \to L^p([0,1]), \quad \boldsymbol{\mu}(A) = \chi_A;$$

$\boldsymbol{\mu}$ is σ-additive if $p \in [1, \infty)$ and additive if $p = \infty$.

Much of the terminology used in the case of vector capacities parallels that of real-valued capacities (presented in Sects. 2.1 and 2.2).

A vector capacity $\boldsymbol{\mu} : \mathcal{A} \to E_+$ is called *strongly subadditive* if

$$\boldsymbol{\mu}(A \cup B) + \boldsymbol{\mu}(A \cap B) \le \boldsymbol{\mu}(A) + \boldsymbol{\mu}(B) \quad \text{for all } A, B \in \mathcal{A} \tag{4.4}$$

and it is called *strongly superadditive* when the inequality (4.4) works in the reversed way. Every additive measure taking values in E_+ is both strongly subadditive and strongly superadditive.

Notice that strong subadditivity implies *subadditivity* i.e.,

$$\boldsymbol{\mu}(A \cup B) \leq \boldsymbol{\mu}(A) + \boldsymbol{\mu}(B) \quad \text{for all } A, B \in \mathcal{A}. \tag{4.5}$$

A vector capacity $\boldsymbol{\mu} : \mathcal{A} \to E_+$ is called *lower continuous*(or continuous by ascending sequences) if

$$\lim_{n\to\infty} \boldsymbol{\mu}(A_n) = \boldsymbol{\mu}\left(\bigcup\nolimits_{n=1}^{\infty} A_n\right)$$

for every nondecreasing sequence $(A_n)_n$ of sets in $\mathcal{A}$ such that $\bigcup_{n=1}^{\infty} A_n \in \mathcal{A}$; $\boldsymbol{\mu}$ is called *upper continuous*(or continuous by descending sequences) if

$$\lim_{n\to\infty} \boldsymbol{\mu}(A_n) = \boldsymbol{\mu}\left(\bigcap\nolimits_{n=1}^{\infty} A_n\right)$$

for every nonincreasing sequence $(A_n)_n$ of sets in $\mathcal{A}$ such that $\bigcap_{n=1}^{\infty} A_n \in \mathcal{A}$. If $\boldsymbol{\mu}$ is an additive capacity defined on a σ-algebra, then its upper/lower continuity is equivalent to the property of σ-additivity.

To each vector capacity $\boldsymbol{\mu}$ defined on the algebra $\mathcal{A}$, one can attach a new vector capacity $\overline{\boldsymbol{\mu}}$, the *dual* of $\boldsymbol{\mu}$, which is defined by the formula

$$\overline{\boldsymbol{\mu}}(A) = \boldsymbol{\mu}(\Omega) - \boldsymbol{\mu}(\Omega \setminus A).$$

Notice that $\overline{(\overline{\boldsymbol{\mu}})} = \boldsymbol{\mu}$ and the dual of a strongly subadditive (strongly superadditive) vector capacity is a strongly superadditive (strongly subadditive) vector capacity. Also, the dual of a lower continuous (upper continuous) capacity is an upper continuous (lower continuous) capacity.

Example 4.2.2 (Distorted Measures in the Vector Case) There are several standard procedures to attach to a σ-additive vector measure $\boldsymbol{\mu} : \Sigma \to E_+$ certain not necessarily additive capacities. So is the case of *distorted measures,* $\boldsymbol{\nu}(A) = T(\boldsymbol{\mu}(A))$, obtained from $\boldsymbol{\mu}$ by applying to it a continuous nondecreasing distortion $T : [0, \boldsymbol{\mu}(\Omega)] \to [0, \boldsymbol{\nu}(\Omega)]$ such that $T(0) = 0$ and $T(\boldsymbol{\mu}(\Omega)) = \boldsymbol{\nu}(\Omega)$. The vector capacities ν so obtained are both upper and lower continuous.

For example, this is the case when $E = \operatorname{Sym}(n, \mathbb{R})$, $\boldsymbol{\mu}(\Omega) = \mathrm{I}$ (the identity of $\mathbb{R}^N$), and $T : [0, \mathrm{I}] \to [0, \mathrm{I}]$ is the distortion defined by the formula $T(A) = A^2$.

Some aspects (but not all) of the theory of vector capacities are straightforward consequences of the theory of real-valued capacities.

Lemma 4.2.3 *A set function $\mu : \mathcal{A} \to E_+$ is a strongly subadditive (strongly superadditive, lower continuous, upper continuous, etc.) vector capacity if and only if $x^* \circ \mu$ is a strongly subadditive (strongly superadditive, lower continuous, upper continuous, etc.) $\mathbb{R}_+$-valued capacity whenever $x^* \in E_+^*$.*

When $E = \mathbb{R}^N$, this assertion can be easily reformulated in terms of the components $\mu_k = \mathrm{pr}_k \circ \mu$ of μ.

Lemma 4.2.3 is a consequence of the assertion (iii) of Lemma 1.2.2.

For the convenience of the reader, we next recall some basic results concerning the Bochner theory of integration of vector-valued functions. For more, see the excellent books by Diestel and Uhl [52] and Yosida [190].

Definition 4.2.4 A function $\psi : \mathbb{R} \to E$ is *Bochner integrable* with respect to the Lebesgue measure on $\mathbb{R}$ if there exists a sequence of simple functions $\psi_n : \mathbb{R} \to E$ such that

$$\lim_{n\to\infty} \psi_n(t) = \psi(t) \text{ almost everywhere and } \int_{\mathbb{R}} \|\psi - \psi_n\| \,\mathrm{d}t \to 0.$$

In this case, the *Bochner integral* of ψ is defined by

$$\int_{\mathbb{R}} \psi \,\mathrm{d}t = \lim_{n\to\infty} \int_{\mathbb{R}} \psi_n \,\mathrm{d}t.$$

Notice that if $T : E \to F$ is a bounded linear operator, then

$$T\left(\int_{\mathbb{R}} \psi \,\mathrm{d}t\right) = \int_{\mathbb{R}} T \circ \psi \,\mathrm{d}t; \tag{4.6}$$

in particular, this holds for all functionals $x^* \in E^*$.

In the context of Bochner integral, the functions $\psi : \mathbb{R} \to E$, which are the limit almost everywhere of a sequence of simple functions are usually called *strongly measurable* to distinguish them from the weakly measurable functions $\varphi : \mathbb{R} \to E$ which are the functions such that $x^* \circ \varphi$ is measurable (in the sense of Definition 2.3.1), whenever $x^* \in E^*$. Clearly, the strongly measurable functions are also weakly measurable. The following result clarifies when the converse holds.

Theorem 4.2.5 (Pettis Measurability Theorem) *A function f defined on a measure space (Ω, Σ, μ) and taking values in a separable Banach space E is strongly measurable if and only if it is weakly measurable.*

For details, see Diestel and Uhl [52], Theorem 2, p. 42.

Theorem 4.2.6 (Bochner's Criterion of Integrability) *A strongly measurable function* $\psi : \mathbb{R} \to E$ *is* Bochner integrable *with respect to the Lebesgue measure on* $\mathbb{R}$ *if and only if*

$$\int_{\mathbb{R}} \|\psi(t)\| \, \mathrm{d}t < \infty;$$

in this case

$$\left\| \int_{\mathbb{R}} \psi(t) \, \mathrm{d}t \right\| \leq \int_{\mathbb{R}} \|\psi(t)\| \, \mathrm{d}t.$$

For details, see Diestel and Uhl [52], Theorem 2, p. 45.

We can now introduce the main notion of interest in this section.

Definition 4.2.7 A measurable function $f : \Omega \to \mathbb{R}$ is called Choquet–Bochner integrable with respect to the vector capacity $\boldsymbol{\mu} : \Sigma \to E_+$ if for every A in Σ the functions

$$t \to \boldsymbol{\mu}(\{x \in A : f(x) \geq t\})$$

and

$$t \to \boldsymbol{\mu}(\{x \in A : f(x) \geq t\}) - \boldsymbol{\mu}(A)$$

are Bochner integrable respectively on $[0, \infty)$ and $(-\infty, 0]$. Under these circumstances, the Choquet–Bochner integral over A is defined by the formula

$$\begin{aligned}(\mathrm{CB}) \int_A f \, \mathrm{d}\boldsymbol{\mu} = &\int_0^{+\infty} \boldsymbol{\mu}(\{x \in A : f(x) \geq t\}) \, \mathrm{d}t \\ &+ \int_{-\infty}^0 [\boldsymbol{\mu}(\{x \in A : f(x) \geq t\}) - \boldsymbol{\mu}(A)] \, \mathrm{d}t.\end{aligned}$$

If $f \geq 0$, then the last integral in the above formula is 0.

Remark 4.2.8 According to the formula (4.6), if f is Choquet–Bochner integrable, then

$$x^* \left((\mathrm{CB}) \int_A f \, \mathrm{d}\boldsymbol{\mu} \right) = (\mathrm{C}) \int_A f \, \mathrm{d}(x^* \circ \boldsymbol{\mu}), \tag{4.7}$$

for every positive linear functional $x^* \in E^*$; notice that the integral in the right-hand side is a usual Choquet integral. Combining this fact with Lemma *2.4.2*, we infer that the inequality sign $\geq$ in the formulas defining the Choquet–Bochner integral can be replaced by $>$.

By combining Bochner's criterion of integrability and Pettis' Theorem of measurability, one obtains the Choquet–Bochner integrability of bounded measurable functions.

Lemma 4.2.9 *If $f : \Omega \to \mathbb{R}$ is a bounded measurable function, then it is Choquet–Bochner integrable with respect to every vector capacity $\boldsymbol{\mu}$ taking values in the positive cone of a separable ordered Banach space E.*

Proof Suppose that $|f| \leq M$, where $M > 0$ is a positive real number and choose arbitrarily a set $A \in \mathcal{A}$. Then the function $\boldsymbol{\varphi}(t) = \boldsymbol{\mu}(\{x \in A : f(x) \geq t\})$ is positive and decreasing on $\mathbb{R}_+$ and vanishes for $t > M$. Thus, φ is weakly measurable. Since E is separable, it follows from Theorem 4.2.5 that it is also strongly measurable. On the other hand,

$$\int_0^\infty \|\boldsymbol{\varphi}(t)\| \, \mathrm{d}t = \int_0^M \|\boldsymbol{\varphi}(t)\| \, \mathrm{d}t \leq M \, \|\boldsymbol{\mu}(\Omega)\| < \infty.$$

By Bochner's Criterion of Integrability, we infer that the function $\boldsymbol{\varphi}$ is Bochner integrable on the interval $[0, \infty)$. In a similar manner, one can show that the function $\boldsymbol{\mu}(\{x \in A : f(x) \geq t\}) - \boldsymbol{\mu}(A)$ is Bochner integrable on the interval $(-\infty, 0]$. As a consequence, the function f is Choquet–Bochner integrable with respect to the vector capacity $\boldsymbol{\mu}$. ■

The next proposition collects a number of simple (but important) properties of the Choquet–Bochner integral.

Proposition 4.2.10 *Suppose that $\mathcal{A}$ is an algebra of subsets of a set Ω, E is a separable ordered Banach space and $\boldsymbol{\mu} : \mathcal{A} \to E_+$ is a vector capacity. Then, for all measurable bounded functions $f, g \in \mathcal{M}_b(\Omega, \mathcal{A})$ and all $A \in \mathcal{A}$, the following assertions hold:*

$$f \geq 0 \textit{ implies } (\mathrm{CB})\int_A f \, \mathrm{d}\boldsymbol{\mu} \geq 0 \; (\textit{positivity})$$

$$f \leq g \textit{ implies } (\mathrm{CB})\int_A f \, \mathrm{d}\boldsymbol{\mu} \leq (\mathrm{CB})\int_A g \, \mathrm{d}\boldsymbol{\mu} \quad (\textit{monotonicity})$$

$$(\mathrm{CB})\int_A af \, \mathrm{d}\boldsymbol{\mu} = a \cdot (\mathrm{CB})\int_A f \, \mathrm{d}\boldsymbol{\mu} \quad \textit{for all } a \geq 0 \quad (\textit{positive homogeneity})$$

$$(\mathrm{CB})\int_A 1 \cdot \mathrm{d}\boldsymbol{\mu} = \boldsymbol{\mu}(A) \quad (\textit{calibration})$$

$$(\text{CB})\int_A (-f)\,\mathrm{d}\mu = -(\text{CB})\int_A f\,\mathrm{d}\overline{\mu} \quad (\textit{asymmetry}).$$

Proof Using the formula (4.7), all the above assertions reduce to the case of scalar capacities, already covered by Proposition 2.4.5. ■

The property of *invariance to translation* i.e.,

$$(\text{CB})\int_\Omega (f+\alpha)\,\mathrm{d}\mu = (\text{CB})\int_\Omega f\,\mathrm{d}\mu + \alpha\cdot\mu(\Omega)$$

for all $\alpha \in \mathbb{R}$ and all functions $f \in \mathcal{M}_b(\Omega, \mathcal{A})$, can be established as in the case of Choquet integral; see formula (2.13).

Theorem 4.2.11 (Comonotonic Additivity) *In general, the Choquet–Bochner integral is not additive but if f and g are comonotonic functions such that f, g and $f+g$ belong to $\mathcal{M}_b(\Omega, \mathcal{A})$, then*

$$(\text{CB})\int_A (f+g)\,\mathrm{d}\mu = (\text{CB})\int_A f\,\mathrm{d}\mu + (\text{CB})\int_A g\,\mathrm{d}\mu.$$

In particular, the Choquet–Bochner integral is translation invariant, that is,

$$(\text{CB})\int_A (f+c)\,\mathrm{d}\mu = (\text{CB})\int_A f\,\mathrm{d}\mu + c\mu(A),$$

for all measurable bounded functions f, all sets $A \in \mathcal{A}$ and all numbers $c \in \mathbb{R}$.

Proof Using the formula (4.7), we may restrict ourselves to the case of scalar capacities, already covered by Theorem 2.5.4. ■

Corollary 4.2.12 *The equality*

$$(\text{CB})\int_A (\alpha f + c)\mathrm{d}\mu = \alpha\left((\text{CB})\int_A f\mathrm{d}\mu\right) + c\cdot\mu(A),$$

holds for all measurable bounded functions f, all sets $A \in \mathcal{A}$ and all numbers $\alpha \in \mathbb{R}_+$ and $c \in \mathbb{R}$.

The next result describes how the special properties of a vector capacity transfer to the Choquet–Bochner integral.

Theorem 4.2.13 *Suppose that E is a separable ordered Banach space and $\mu : \mathcal{A} \to E_+$ is a vector capacity.*

(i) *If μ is an upper continuous capacity, then the Choquet–Bochner integral is an upper continuous operator, i.e.,*

$$\lim_{n\to\infty} \left\| (\mathrm{CB}) \int_A f_n \, \mathrm{d}\mu - (\mathrm{CB}) \int_A f \, \mathrm{d}\mu \right\| = 0,$$

whenever $A \in \mathcal{A}$ and $(f_n)_n$ is a nonincreasing sequence of measurable bounded functions that converges pointwise to the measurable bounded function f.

(ii) *If μ is a lower continuous capacity, then the Choquet–Bochner integral is lower continuous, i.e.,*

$$\lim_{n\to\infty} \left\| (\mathrm{CB}) \int_A f_n \, \mathrm{d}\mu - (\mathrm{CB}) \int_A f \, \mathrm{d}\mu \right\| = 0,$$

whenever $A \in \mathcal{A}$ and $(f_n)_n$ is a nondecreasing sequence of measurable bounded functions that converges pointwise to the measurable bounded function f.

Proof (i) Since μ is an upper continuous capacity and $(f_n)_n$ is a nonincreasing sequence of measurable functions that converges pointwise to the measurable function f it follows that

$$\mu\left(\{x \in A : f_n(x) \geq t\}\right) \searrow \mu\left(\{x \in A : f(x) \geq t\}\right)$$

in the norm topology. Taking into account the property of monotonicity of the Choquet–Bochner integral (already noticed in Proposition 4.2.10), we have

$$(\mathrm{CB}) \int_A f_n \, \mathrm{d}\mu \geq (\mathrm{CB}) \int_A f_n \, \mathrm{d}\mu \geq \cdots \geq (\mathrm{CB}) \int_A f \, \mathrm{d}\mu,$$

so by the Theorem of monotone convergence for the Lebesgue integral, it follows that

$$x^* \left((\mathrm{CB}) \int_A f_n \, \mathrm{d}\mu \right) = (\mathrm{C}) \int_A f_n \, \mathrm{d}\mu^* \to (\mathrm{C}) \int_A f \, \mathrm{d}\mu^* = x^* \left((\mathrm{CB}) \int_A f \, \mathrm{d}\mu \right)$$

for all $x^* \in E_+^*$. The conclusion of the assertion (i) is now a direct consequence of the generalized Dini's lemma (see Lemma 1.2.2 (v)).

(ii) The argument is similar to that used for the assertion (i). ∎

The property of subadditivity of the Choquet–Bochner integral makes the objective of the following result.

Theorem 4.2.14 (The Subadditivity Theorem) *Suppose that E is a separable ordered Banach space and $\boldsymbol{\mu} : \mathcal{A} \to E_+$ is a vector capacity. Then the following assertions are equivalent:*

(i) *$\boldsymbol{\mu}$ is a strongly subadditive vector capacity, i.e.,*

$$\boldsymbol{\mu}(A \cup B) + \boldsymbol{\mu}(A \cap B) \leq \boldsymbol{\mu}(A) + \boldsymbol{\mu}(B)$$

for all $A, B \in \mathcal{A}$;

(ii) *the Choquet–Bochner integral operator associated with $\boldsymbol{\mu}$ is subadditive, i.e.,*

$$(\mathrm{C}) \int_\Omega (f+g)\, \mathrm{d}\boldsymbol{\mu} \leq (\mathrm{C}) \int_\Omega f\, \mathrm{d}\boldsymbol{\mu} + (\mathrm{C}) \int_\Omega g\, \mathrm{d}\boldsymbol{\mu}$$

for all functions $f, g \in \mathcal{M}_b(\Omega, \mathcal{A})$ such that $f + g \in \mathcal{M}_b(\Omega, \mathcal{A})$;

(iii) *the Choquet–Bochner integral operator associated with $\boldsymbol{\mu}$ is a submodular operator in the sense that*

$$(\mathrm{C}) \int_\Omega f \vee g\, \mathrm{d}\boldsymbol{\mu} + (\mathrm{C}) \int_\Omega f \wedge g\, \mathrm{d}\boldsymbol{\mu} \leq (\mathrm{C}) \int_\Omega f\, \mathrm{d}\boldsymbol{\mu} + (\mathrm{C}) \int_\Omega g\, \mathrm{d}\boldsymbol{\mu}$$

for all functions $f, g \in \mathcal{M}_b(\Omega, \mathcal{A})$;

(iv) *the Choquet–Bochner integral operator associated with $\boldsymbol{\mu}$ is a convex operator in the sense that*

$$(\mathrm{C}) \int_\Omega ((1-\alpha)f + \alpha g)\, \mathrm{d}\boldsymbol{\mu} \leq (1-\alpha) \cdot (\mathrm{C}) \int_\Omega f\, \mathrm{d}\boldsymbol{\mu} + \alpha \cdot (\mathrm{C}) \int_\Omega g\, \mathrm{d}\boldsymbol{\mu}$$

for all $\alpha \in [0, 1]$ and all functions $f, g \in \mathcal{M}_b(\Omega, \mathcal{A})$ such that $(1-\alpha)f + \alpha g \in \mathcal{M}_b(\Omega, \mathcal{A})$.

Proof (i) $\Rightarrow$ (ii) According to Lemma 4.2.3, if $\boldsymbol{\mu}$ is strongly subadditive, then the real-valued capacity $\mu^* = x^* \circ \boldsymbol{\mu}$ is also strongly subadditive, whenever $x^* \in E_+^*$. Then

$$(\mathrm{C}) \int_A (f+g)\, \mathrm{d}\mu^* \leq (\mathrm{C}) \int_A f\, \mathrm{d}\mu^* + (\mathrm{C}) \int_A g\, \mathrm{d}\mu^*,$$

as a consequence of Theorem 2.6.1. The assertion (ii) is now a direct consequence of Lemma 1.2.1 (c) and the formula (4.6).

(ii) $\Rightarrow$ (i) For every two sets $A, B \in \mathcal{A}$, the functions $\chi_{A \cup B}$ and $\chi_{A \cap B}$ are comonotonic, so that, according to Theorem 4.2.11, we have

$$\begin{aligned}\mu(A\cup B)+\mu(A\cap B)&=(\mathrm{C})\int_\Omega \chi_{A\cup B}\,\mathrm{d}\mu+(\mathrm{C})\int_\Omega \chi_{A\cap B}\,\mathrm{d}\mu\\&=(\mathrm{C})\int_\Omega(\chi_{A\cup B}+\chi_{A\cap B})\,\mathrm{d}\mu=(\mathrm{C})\int_\Omega(\chi_A+\chi_B)\,\mathrm{d}\mu\\&\le(\mathrm{C})\int_\Omega \chi_A\,\mathrm{d}\mu+(\mathrm{C})\int_\Omega \chi_B\,\mathrm{d}\mu\\&=\mu(A)+\mu(B).\end{aligned}$$

(i) $\Rightarrow$ (iii) Since $\mathcal{A}$ is an algebra, both functions $\inf\{f,g\}$ and $\sup\{f,g\}$ are measurable. The fact that μ is strongly subadditive implies

$$\begin{aligned}&\mu(\{x:\sup\{f,g\}(x)\ge t\})+\mu(\{x:\inf\{f,g\}(x)\ge t\})\\&\quad=\mu(\{x:f(x)\ge t\}\cup\{x:g(x)\ge t\})+\mu(\{x:f(x)\ge t\}\cap\{x:g(x)\ge t\})\\&\qquad\le\mu(\{x:f(x)\ge t\})+\mu(\{x:g(x)\ge t\}),\end{aligned}$$

and the same works when μ is replaced by $\mu^*=x^*\circ\mu$, where $x^*\in E_+^*$ is arbitrarily fixed. Integrating side by side the last inequality, it follows that

$$(\mathrm{C})\int_A\sup\{f,g\}\,\mathrm{d}\mu^*+(\mathrm{C})\int_A\inf\{f,g\}\,\mathrm{d}\mu^*\le(\mathrm{C})\int_A f\,\mathrm{d}\mu^*+(\mathrm{C})\int_A g\,\mathrm{d}\mu^*,$$

which yields (via Lemma 1.2.2 (iii) and the formula (4.6)) the submodularity of the Choquet–Bochner integral.

For (iii) $\Rightarrow$ (i), notice that for $f=\chi_A$ and $g=\chi_B$, we have $f\vee g=\chi_{A\cup B}$ and $f\wedge g=\chi_{A\cap B}$, so in this case the assertion (iii) coincides with the condition of strong subadditivity of μ.

Finally, the equivalence of the assertions (ii) and (iv) follows from the positive homogeneity of the Choquet–Bochner integral. ■

Corollary 4.2.15 (The Lipschitz Continuity of the Choquet–Bochner Integral) *If E is a separable ordered Banach space and μ is a strongly subadditive capacity, then the associated Choquet–Bochner integral is a sublinear and monotone increasing operator such that*

$$\left\|(\mathrm{CB})\int_A f\,\mathrm{d}\mu-(\mathrm{CB})\int_A g\,\mathrm{d}\mu\right\|\le\|\mu(\Omega)\|\sup_{x\in\Omega}|f(x)-g(x)|$$

for all $f,g\in\mathcal{M}_b(\Omega,\mathcal{A})$ and $A\in\mathcal{A}$.

Proof In order to prove the inequality, notice that

$$\begin{aligned}(\mathrm{CB})\int_A f\,\mathrm{d}\boldsymbol{\mu} &= (\mathrm{CB})\int_A (f-g+g)\,\mathrm{d}\boldsymbol{\mu}\\ &\leq (\mathrm{CB})\int_A (f-g)\,\mathrm{d}\boldsymbol{\mu} + (\mathrm{CB})\int_A g\,\mathrm{d}\boldsymbol{\mu},\end{aligned}$$

whence

$$\begin{aligned}(\mathrm{CB})\int_A f\,\mathrm{d}\boldsymbol{\mu} - (\mathrm{CB})\int_A g\,\mathrm{d}\boldsymbol{\mu} &\leq (\mathrm{CB})\int_A (f-g)\,\mathrm{d}\boldsymbol{\mu}\\ &\leq (\mathrm{CB})\int_A |f-g|\,\mathrm{d}\boldsymbol{\mu}\\ &\leq \sup_{x\in\Omega}|f(x)-g(x)|\cdot\boldsymbol{\mu}(\Omega).\end{aligned}$$

This implies that

$$\left|x^*\left[(\mathrm{CB})\int_A f\,\mathrm{d}\boldsymbol{\mu} - (\mathrm{CB})\int_A g\,\mathrm{d}\boldsymbol{\mu}\right]\right| \leq \sup_{x\in\Omega}|f(x)-g(x)|\cdot x^*\left(\boldsymbol{\mu}(\Omega)\right).$$

for every functional $x^* \in E_+^*$ with $\|x^*\| \leq 1$. Taking the supremum over x^* in the both sides, we deduce from Lemma 1.2.2 (iv) that

$$\left\|(\mathrm{CB})\int_A f\,\mathrm{d}\boldsymbol{\mu} - (\mathrm{CB})\int_A g\,\mathrm{d}\boldsymbol{\mu}\right\| \leq \|\boldsymbol{\mu}(\Omega)\|\sup_{x\in\Omega}|f(x)-g(x)|$$

and the proof is complete. ■

Corollary 4.2.16 (The Modulus Inequality) *Suppose that E is a separable Banach lattice and $\boldsymbol{\mu} : \mathcal{A} \to E_+$ is a strongly subadditive vector capacity. Then the following integral analog of the modulus inequality holds:*

$$\left|(\mathrm{CB})\int_A f\,\mathrm{d}\boldsymbol{\mu} - (\mathrm{CB})\int_A g\,\mathrm{d}\boldsymbol{\mu}\right| \leq (\mathrm{CB})\int_A |f-g|\,\mathrm{d}\boldsymbol{\mu}$$

whenever f, g, $f-g$ and $g-f$ belong to $\mathcal{M}_b(\Omega,\mathcal{A})$ and $A \in \mathcal{A}$. In particular,

$$\left|(\mathrm{CB})\int_A f\,\mathrm{d}\boldsymbol{\mu}\right| \leq (\mathrm{CB})\int_A |f|\,\mathrm{d}\boldsymbol{\mu},$$

whenever f and $-f$ belong to $\mathcal{M}_b(\Omega,\mathcal{A})$ and $A \in \mathcal{A}$.

Proof Proceeding as in the proof of Corollary 4.2.15, notice that

$$\pm\left((\mathrm{CB})\int_A f\,\mathrm{d}\mu - (\mathrm{CB})\int_A g\,\mathrm{d}\mu\right) \leq (\mathrm{CB})\int_A |f-g|\,\mathrm{d}\mu,$$

a fact that is equivalent to the inequality stated by Corollary 4.2.16. ■

4.3 The Representation Theorem: The Vector-Valued Case

As in the case of Choquet functionals, all comonotonic additive and monotone increasing operators $T : \mathcal{M}_b(\Omega, \mathcal{A}) \to E$ with a separable range can be represented as Choquet–Bochner integrals with respect to suitable vector-valued capacities. This fact, which extends Schmeidler's representation Theorem 2.8.1, can be established in a similar way. Recall here the basic ingredients.

Lemma 4.3.1 *Let $\mathcal{A}$ be an algebra of subsets of an abstract set Ω and E be an ordered Banach space. Then every comonotonic additive and monotone increasing operator $T : \mathcal{M}_b(\Omega, \mathcal{A}) \to E$ is positively homogeneous.*

The proof follows verbatim the proof of Lemma 2.8.3.

Lemma 4.3.2 *Under the hypotheses of Lemma* 4.3.1*, every translation invariant and monotone increasing operator $T : \mathcal{M}_b(\Omega, \mathcal{A}) \to E$ is Lipschitz continuous.*

Proof Without loss of generality, we may assume that E is regularly ordered. See Lemma 1.2.1. Given $f, g \in \mathcal{M}_b(\Omega, \mathcal{A})$, one can choose a decreasing sequence $(\alpha_n)_n$ of positive numbers such that $\alpha_n \downarrow \|f-g\|$. Then $f \leq g + |f-g| \leq g + \alpha_n \cdot 1$, which implies

$$T(f) \leq T(g + \alpha_n \cdot 1) = T(g) + \alpha_n T(1),$$

due to the properties of monotonicity and translation invariance of T. Since f and g can be interchanged, we are led to the fact that $\pm\,(T(f) - T(g)) \leq T(1) \cdot \alpha_n$ for all n. Since E is regularly ordered, we infer that

$$\|T(f) - T(g)\| \leq \|T(1)\| \cdot \alpha_n.$$

Letting $n \to \infty$, we conclude that

$$\|T(f) - T(g)\| \leq \|T(1)\| \cdot \|f-g\|.$$

■

Now we can state the following generalization of Schmeidler's Representation Theorem 2.8.1.

Theorem 4.3.3 *Let $\mathcal{A}$ be an algebra of subsets of an abstract set Ω, and let E be a separable ordered Banach space. Then every comonotonic additive and monotone increasing operator $T : \mathcal{M}_b(\Omega, \mathcal{A}) \to E$ can be represented as the restriction of the Choquet–Bochner integral to $\mathcal{M}_b(\Omega, \mathcal{A})$, with respect to the vector-valued capacity $\boldsymbol{\mu} : \mathcal{A} \to [0, T(1_\Omega)]$ defined by $\boldsymbol{\mu}(A) = T(\chi_A)$ for every $A \in \mathcal{A}$. In other words,*

$$T(f) = (\mathrm{CB}) \int_\Omega f \, \mathrm{d}\boldsymbol{\mu}$$
$$= \int_0^{+\infty} \boldsymbol{\mu}(\{x : f(x) \geq t\}) \, \mathrm{d}t + \int_{-\infty}^0 [\boldsymbol{\mu}(\{x : f(x) \geq t\}) - T(1_\Omega)] \, \mathrm{d}t$$

for every $f \in \mathcal{M}_b(\Omega, \mathcal{A})$.

Moreover, if T is in addition upper continuous (respectively subadditive), then $\boldsymbol{\mu}$ is upper continuous (respectively strongly subadditive).

Proof The fact that $\boldsymbol{\mu}$ is a capacity is obvious. According to Corollary 4.2.15, combined with Lemma 2.8.4 and the property of translation invariance of T, it suffices to prove that the formula stated in Theorem 4.3.3 holds for the positive simple functions. Every such function f admits a representation of the form

$$f = \sum\nolimits_{i=1}^n \alpha_i \chi_{A_i}$$

with $0 \leq \alpha_n \leq \cdots \leq \alpha_1$ and the sets $A_1, \ldots, A_n \in \mathcal{A}$ are pairwise disjoint and nonempty. According to Example 2.4.9,

$$(\mathrm{CB}) \int_\Omega f \, \mathrm{d}\boldsymbol{\mu} = \int_0^{+\infty} \boldsymbol{\mu}(\{x : f(x) \geq t\}) \, \mathrm{d}t$$
$$= \sum_{i=1}^n (\alpha_i - \alpha_{i+1}) \boldsymbol{\mu}\left(\bigcup\nolimits_{j=1}^i A_j\right),$$

where $\alpha_{n+1} = 0$ by definition. It remains to prove that also

$$T(f) = \sum_{i=1}^n (\alpha_i - \alpha_{i+1}) \boldsymbol{\mu}\left(\bigcup\nolimits_{j=1}^i A_j\right), \tag{4.8}$$

which can be done by induction over n. For $n = 1$, $f = \alpha_1 \chi_{A_1}$ and in this case

$$T(\alpha_1 \chi_{A_1}) = \alpha_1 T(\chi_{A_1}) = \alpha_1 \boldsymbol{\mu}(A_1)$$

since T is positively homogeneous. Assuming the formula 4.8 true for all simple functions f with $k < n$ summands, we will prove that the same happens for n summands. Indeed, we can split f as

$$f = g + h,$$

where

$$g = \sum\nolimits_{k=1}^{n-1} (\alpha_k - \alpha_n)\, \chi_{A_k} \text{ and } h = \alpha_n \chi_{\cup_{k=1}^{n} A_k}.$$

According to the induction hypothesis,

$$\begin{aligned} T(g) &= \sum_{i=1}^{n-1} ((\alpha_i - \alpha_n) - (\alpha_{i+1} - \alpha_n))\mu \left(\bigcup\nolimits_{j=1}^{i} A_j \right) \\ &= \sum_{i=1}^{n-1} (\alpha_i - \alpha_{i+1})\mu \left(\bigcup\nolimits_{j=1}^{i} A_j \right). \end{aligned}$$

and

$$T(h) = \alpha_n \mu \left(\bigcup\nolimits_{k=1}^{n} A_k \right).$$

Since g and h are comonotonic, it follows that $T(f) = T(g+h) = T(g) + T(h)$ and the proof is done.

The second part of Theorem 4.3.3 follows from Theorems 4.2.13 (i) and 4.2.14. ■

We pass now to the integral representation of Choquet-type operators defined on spaces of continuous functions and taking values in an ordered Banach space. This will be derived from Corollary 2.9.2 and the following couple of auxiliary lemmas.

Lemma 4.3.4 *Suppose that K is a compact Hausdorff space and E is an ordered Banach space. Then:*

(i) *Every comonotonic additive and monotone increasing operator T mapping $C(K)$ into E is positively homogeneous;*
(ii) *Every translation invariant and monotone increasing operator T mapping $C(K)$ into E is Lipschitz continuous and lower continuous.*

Proof (i) See the proof of Lemma 2.8.3.

(ii) The fact that T is a Lipschitz operator can be shown as in the case of Lemma 4.3.2. To prove that T is also lower continuous, let us consider an arbitrary increasing sequence

$(f_n)_n$ of functions in $C(K)$ that converges pointwise to a function $f \in C(K)$. Since T is monotone increasing, for every $x^* \in E_+^*$, we have

$$\lim_{n\to\infty} x^*(T(f_n)) \leq x^*(T(f)).$$

According to Dini's lemma, the convergence of $(f_n)_n$ to f is uniform, so for $\varepsilon > 0$ arbitrarily fixed there is an integer N, such that $f_n(x) \geq f(x) - \varepsilon 1$, for all $x \in X$ and $n \geq N$, whence

$$\lim_{n\to\infty} x^*(T(f_n)) \geq x^*(T(f - \varepsilon 1)),$$

due to the fact that T is monotone increasing. Since T is translation invariant, it follows that

$$T(f) = T(f - \varepsilon 1 + \varepsilon 1) = T(f - \varepsilon 1) + \varepsilon T(1),$$

which implies that

$$x^*(T(f - \varepsilon 1)) = x^*(T(f)) - \varepsilon x^*(T(1)).$$

Therefore,

$$\lim_{n\to\infty} x^*(T(f_n)) \geq x^*(T(f - \varepsilon 1)) = x^*(T(f)) - \varepsilon x^*(T(1))$$

and passing to the limit as $\varepsilon \to 0$, we get $\lim_{n\to\infty} x^*(T(f_n)) \geq x^*(T(f))$. Since the other inequality was already proved, we conclude that $\lim_{n\to\infty} x^*(T(f_n)) = x^*(T(f))$ for every $x^* \in E_+^*$, i.e., $\lim_{n\to\infty} T(f_n) = T(f)$, as $x^* \in E_+^*$ was arbitrarily chosen. See Lemma 1.2.2 (c). ■

Lemma 4.3.5 *Suppose that K is a compact Hausdorff space and E is an ordered Banach space with order continuous norm. Then every monotone increasing, positively homogeneous, and translation invariant operator $T : C(K) \to E$ is weakly compact and upper continuous.*

Proof The weak compactness of T follows from the fact that T maps the closed order intervals in $C(K)$ (that is, all closed balls) into closed order intervals in E and all such intervals are weakly compact in E. See Lemma 1.2.4.

For the property of upper continuity, let $(f_n)_n$ be any nonincreasing sequence of functions in $C(K)$, which converges pointwise to a continuous function f. According to Dini's lemma, the sequence $(f_n)_n$ is convergent to f in the norm topology of $C(K)$. Hence, for each $\varepsilon > 0$ there is an index N such that $f \leq f_n < f + \varepsilon$ for all $n \geq N$.

Because T is monotone increasing, it follows that

$$T(f) \le T(f_n) \le T(f+\varepsilon) = T(f) + \varepsilon \cdot T(1) \text{ for all } n \ge N, \tag{4.9}$$

where 1 is the unit of $C(K)$. Taking into account that E has order continuous norm and $T(f_n) \ge T(f_{n+1}) \ge T(f)$ for all n, it follows that the limit $\lim_{n\to\infty} T(f_n)$ exists in E and $\lim_{n\to\infty} T(f_n) \ge T(f)$. Combining this fact with (4.9), we infer that $T(f) \le \lim_{n\to\infty} T(f_n) \le T(f) + \varepsilon T(1)$. Since $\varepsilon > 0$ was arbitrarily chosen, we conclude that $\lim_{n\to\infty} T(f_n) = T(f)$. ∎

With this preparation, one can now formulate the following vector analog of Theorem 2.9.1.

Theorem 4.3.6 *Let K be a compact metric space (with metric d) and E be a separable ordered Banach space with order continuous norm. Then for every Choquet-type operator $T : C(K) \to E$ with $T(1)$ a strictly positive element, there exists a unique upper continuous vector capacity $\boldsymbol{\mu} : \mathcal{B}(K) \to E_+$ such that*

$$T(f) = (\mathrm{CB}) \int_K f \, \mathrm{d}\boldsymbol{\mu} \quad \textit{for all } f \in C(K). \tag{4.10}$$

Moreover, $\boldsymbol{\mu}$ verifies the following two properties of regularity:

(IR) $\boldsymbol{\mu}(A) = \sup\{\boldsymbol{\mu}(C) : C$ *compact,* $C \subset A\}$ *for all* $A \in \mathcal{B}(K)$;
(OR) $\boldsymbol{\mu}(C) = \inf\{\boldsymbol{\mu}(O) : O$ *open,* $O \supset C\}$ *for all closed sets* $C \subset K$.

Proof According to Lemmas 4.3.4 and 4.3.4, the operator T is also positively homogeneous, Lipschitz continuous, weakly compact as well as lower and upper continuous.

The characteristic function χ_C of each closed set $C \subset K$ is the pointwise limit of a nonincreasing sequence $(f_n^{(C)})_n$ of continuous and nonnegative functions. Indeed, according to Urysohn's lemma in Real Analysis (see Hewitt and Stromberg [98], Theorem 6.80, p. 75), for every $n \in \mathbb{N}$, there exists a continuous function $h_n^{(C)} : K \to [0, 1]$ such that $h_n^{(C)} = 1$ on C and $h_n^{(C)} = 0$ on the complement of $\{x : d(x, C) < 1/n\}$ in K. Then the sequence

$$f_1^{(C)} = h_1^{(C)},\ f_2^{(C)} = h_1^{(C)} \wedge h_2^{(C)}, \ldots, f_n^{(C)} = h_1^{(C)} \wedge \cdots \wedge h_n^{(C)}, \ldots$$

verifies the aforementioned properties.

Since T is monotone increasing, the sequence $(T(f_n^{(C)}))_n$ is also nonincreasing and bounded from below by 0, which implies (due to the order continuity of the norm of E), that it is also convergent in the norm topology of E. This allows us to define $\boldsymbol{\mu}$ on the lattice $\mathcal{K}$, of all closed sets $C \subset K$, via the formula

$$\mu(C) = \lim_{n\to\infty} T(f_n^{(C)}). \tag{4.11}$$

The definition of $\mu(C)$ is independent of the particular sequence $(f_n^{(C)})_n$ with the aforementioned properties. Indeed, if $(g_n^{(C)})_n$ is another such sequence, fix a positive integer m, and infer from Lemma 2.8.4 (a) that $\lim_{n\to\infty} T(g_n^{(C)}) \leq T(f_m^{(C)})$. Taking the limit as $m \to \infty$ on the right-hand side, we get

$$\lim_{n\to\infty} T(g_n^{(C)}) \leq \lim_{m\to\infty} T(f_m^{(C)}).$$

Then, interchanging $(f_n^{(C)})_n$ and $(g_n^{(C)})_n$, we conclude that the two limits are equal.

Clearly, the set function $\mu : \mathcal{K} \to E_+$ takes values in the order interval $[0, T(1)]$ and verifies the conditions $(C1)$ and $(C2)$ from the Definition 4.2.1 of vector capacities.

The next goal is the representation formula (4.10). For this, let $x^* \in E_+^*$, $x^* \neq 0$, be arbitrarily fixed and consider the comonotonic additive, upper continuous, and monotone increasing functional

$$I^* = x^* \circ T : C(K) \to \mathbb{R}.$$

It verifies $I^*(1) = x^*(I(1)) > 0$, so by Theorem 2.9.1 there is a unique upper continuous capacity $\nu^* : \mathcal{K} \to [0, I^*(1)]$ such that

$$\left(x^* \circ I\right)(f) = (C)\int_K f\, \mathrm{d}\nu^* \quad \text{for all } f \in C(K).$$

The capacity ν^* is obtained via an approximation process similar to (4.11). Therefore,

$$\nu^*(C) = \lim_{n\to\infty} x^*(I(f_n^{(C)})) = x^*(\lim_{n\to\infty} [I(f_n^{(C)})) = x^*(\mu(C))$$

for all $C \in \mathcal{K}$,, which implies that

$$x^*(T(f)) = (\mathrm{C})\int_K f\, \mathrm{d}(x^* \circ \mu).$$

Since $x^* \in E_+^*$ was arbitrarily fixed, an appeal to Lemma 1.2.1 (c) easily yields the equality

$$T(f) = (\mathrm{CB})\int_K f\, \mathrm{d}\mu.$$

For the second part of Theorem 4.3.6, since μ takes values in an order bounded interval, one can extend it to all Borel subsets of K via the formula

$$\mu(A) = \sup\{\mu(C) : C \text{ closed}, C \subset A\}, \quad A \in \mathcal{B}(K).$$

The fact that the resulting set function μ is a vector capacity is immediate.

This set function μ also verifies the regularity condition (OR). Indeed, given a closed set C, we can consider the sequence of open sets

$$O_n = \{x \in K : d(x, C) < 1/n\}.$$

Clearly, $\mu(C) \leq \mu(O_n) \leq \mu(C_n)$ where

$$C_n = \{x \in X : d(x, C) \leq 1/n\}.$$

Since μ is upper continuous on closed sets, it follows that $\lim_{n\to\infty} \mu(C_n) = \mu(C)$, whence $\lim_{n\to\infty} \mu(O_n) = \mu(C)$. Therefore, μ verifies the regularity condition (OR). The uniqueness of the extension of μ to $\mathcal{B}(K)$ is motivated by the condition (IR). ■

Remark 4.3.7 If the operator T is submodular i.e.,

$$T(\sup\{f, g\}) + T(\inf\{f, g\}) \leq T(f) + T(g) \quad \text{for all } f, g \in C(K),$$

then the vector capacity $\mu : \mathcal{B}(K) \to E_+$ stated by Theorem 4.3.6 is strongly subadditive. This is a consequence of Theorem 4.2.14.

Open Problem 4.3.8 *The classical result due to Bartle, Dunford, and Schwartz* [18] *concerning the integral representation of vector-valued weakly compact operators defined on spaces of continuous functions suggests that* Theorem 4.3.6 *could be improved by removing the condition of order continuity of the norm of E at the cost of weak compactness of T. However, the details remain to be settled.*

4.4 The Case of Operators with Bounded Variation

The representation Theorems 4.3.3 and 4.3.6 can be extended outside the framework of monotone increasing operators, by considering the class of operators with bounded variation. We will discuss here the extension of Theorem 4.3.6, by adapting the argument used by Cerreia-Vioglio, Maccheroni, Marinacci, and Montrucchio [35] in the case of functionals.

In what follows K is a compact metric space and E is a Banach lattice with order continuous norm.

Definition 4.4.1 An operator $T : C(K) \to E$ has *bounded variation* over an order interval $[f, g]$ if

$$\vee_f^g T = \sup \sum_{k=0}^{n} |T(f_k) - T(f_{k-1})| \quad \text{exists in } E, \tag{4.12}$$

the supremum being taken over all finite chains $f = f_0 \leq f_1 \leq \cdots \leq f_n = g$ of functions in the Banach lattice $C(K)$. The operator T has bounded variation (over $C(K)$) if it has bounded variation on every order intervals $[f, g]$ in $C(K)$.

Clearly, if T is monotone increasing, then $\vee_f^g T = T(g) - T(f)$ for all $f \leq g$ in $C(K)$ and thus T has bounded variation.

More generally, every operator $T : C(K) \to E$, which can be represented as a difference $T = T_1 - T_2$ of two monotone increasing operators $T_1, T_2 : C(K) \to E$, has bounded variation. This follows from the order completeness of E and the modulus inequality, which provides an upper bound for the sums appearing in formula (4.12):

$$\sum_{k=0}^{n} |T(f_k) - T(f_{k-1})| \leq T_1(g) - T_1(f) + T_2(g) - T_2(f).$$

The converse also holds. The basic ingredient is the following result.

Lemma 4.4.2 *Suppose that $T : C(K) \to E$ is a positively homogeneous and comonotonic additive operator with bounded variation. Then there exist two positively homogeneous, translation invariant and monotone increasing operators $T_1, T_2 : C(K) \to E$ such that $T = T_1 - T_2$.*

Moreover, if T is upper continuous, then both operators T_1 and T_2 can be chosen to be upper continuous.

Proof The proof is done in the footsteps of Lemma 14 in [35], by noticing first the following four facts:

(a) According to our hypotheses, $T(\alpha f + \beta) = \alpha T(f) + T(\beta)$ for all $f \in C(K)$, $\alpha \in \mathbb{R}_+$ and $\beta \in \mathbb{R}$.
(b) $\vee_0^{f+\alpha} T = \vee_{-\alpha}^{f} T$ for all $f \in C(K)$ and $\alpha \in \mathbb{R}$ with $f + \alpha \geq 0$.

This follows from the definition of the variation.

(c) $\vee_0^{\alpha f} T = \alpha \vee_0^{f} T$ for all $f \in C(K)_+$ and $\alpha \in \mathbb{R}_+$.

Indeed for every $\varepsilon \in E$, $\varepsilon > 0$,, there exists a chain $0 = f_0 \leq f_1 \leq \cdots \leq f_n = f$ such that

$$\sum_{k=0}^{n} |T(f_k) - T(f_{k-1})| \geq \vee_0^f T - \varepsilon.$$

According to fact (a), the chain $0 = \alpha f_0 \leq \alpha f_1 \leq \cdots \leq \alpha f_n = \alpha f$ verifies

$$\vee_0^{\alpha f} T \geq \sum_{k=0}^{n} |T(\alpha f_k) - T(\alpha f_{k-1})| \geq \alpha \vee_0^f T - \alpha\varepsilon.$$

As $\varepsilon > 0$ was arbitrarily fixed, it follows that $\vee_0^{\alpha f} T \geq \alpha \vee_0^f T$. By replacing α to $1/\alpha$ and then f by αf, one obtains the reverse inequality, $\vee_0^{\alpha f} T \leq \alpha \vee_0^f T$.

(d) $\vee_{-\alpha}^f T = \vee_{-\alpha}^0 T + \vee_0^f T = \vee_0^\alpha T + \vee_0^f T$ for all $f \in C(K)_+$ and $\alpha \in \mathbb{R}_+$.

The fact that $\vee_{-\alpha}^f T \geq \vee_{-\alpha}^0 T + \vee_0^f T = \vee_0^\alpha T + \vee_0^f T$ is a direct consequence of the definition of variation. For the other inequality, fix arbitrarily $\varepsilon > 0$ in E and choose a chain $-\alpha = f_0 \leq f_1 \leq \cdots \leq f_n = f$ such that

$$\sum_{k=0}^{n} |T(f_k) - T(f_{k-1})| \geq \vee_{-\alpha}^f T - \varepsilon.$$

Then $-\alpha = -f_0^- \leq -f_1^- \leq \cdots \leq -f_n^- = 0$ and $0 = f_0^+ \leq f_1^+ \leq \cdots \leq f_n^+ = f$. Since all pairs $-f_k^-, f_k^+$ are comonotonic, we have

$$\begin{aligned}
\vee_{-\alpha}^0 T + \vee_0^f T &\geq \sum_{k=0}^{n} \left|T(-f_k^-) - T(-f_{k-1}^-)\right| + \sum_{k=0}^{n} \left|T(f_k^+) - T(f_{k-1}^+)\right| \\
&\geq \sum_{k=0}^{n} \left|T(-f_k^-) - T(-f_{k-1}^-) + T(f_k^+) - T(f_{k-1}^+)\right| \\
&= \sum_{k=0}^{n} |T(f_k) - T(f_{k-1})| \geq \vee_{-\alpha}^f T - \varepsilon
\end{aligned}$$

and it remains to let $\varepsilon \to 0+$.

Now we can proceed to the choice of the operators T_1 and T_2. By definition,

$$T_1(f) = \vee_0^f T \quad \text{if } f \in C(K)_+,$$

while if $f \in C(K)$ is an arbitrary function, one chooses $\alpha \in \mathbb{R}_+$ such that $f + \alpha \geq 0$ and put

$$T_1(f) = \vee_0^{f+\alpha} T - \alpha \vee_0^1 T.$$

The fact that T_1 is well-defined (that is, independent of α) can be proved as follows. Suppose that $\alpha, \beta > 0$ are two numbers such that $f + \alpha \geq 0$ and $f + \beta \geq 0$. Without loss of generality, we may assume that $\alpha < \beta$. Indeed, according to the facts $(b) - (d)$, we have

$$\begin{aligned}\vee_0^{f+\beta} T - \beta \vee_0^1 T &= \vee_0^{f+\alpha+(\beta-\alpha)} T - \beta \vee_0^1 T \overset{\text{fact }(b)}{=} \vee_{-(\beta-\alpha)}^{f+\alpha} T - \beta \vee_0^1 T \\ &\overset{\text{fact }(d)}{=} \vee_{-(\beta-\alpha)}^{0} T + \vee_0^{f+\alpha} T - \beta \vee_0^1 T \\ &\overset{\text{fact }(b)}{=} \vee_0^{\beta-\alpha} T + \vee_0^{f+\alpha} T - \beta \vee_0^1 T \\ &\overset{\text{fact }(c)}{=} (\beta-\alpha) \vee_0^1 T + \vee_0^{f+\alpha} T - \beta \vee_0^1 T \\ &= \vee_0^{f+\alpha} T - \alpha \vee_0^1 T.\end{aligned}$$

By definition,

$$T_2 = T - T_1.$$

Let f, g be two functions in $C(K)$ such that $f \leq g$, and let $\alpha > 0$ such that $f + \alpha \geq 0$. Since T is comonotonic additive and has bounded variation,

$$0 \leq T(g) - T(f) = T(g+\alpha) - T(f+\alpha) \leq \vee_{f+\alpha}^{g+\alpha} T \leq \vee_0^{g+\alpha} T - \vee_0^{f+\alpha} T = T_1(g) - T_1(f),$$

whence we infer that both operators T_1 and T_2 are monotonic.

Due to the fact (c), the operator T_1 is positively homogeneous. Therefore, the same is true for T_2.

For the property of translation invariance, let $f \in C(K)$, $\beta \in \mathbb{R}$ and choose $\alpha > 0$ such that $f + \beta + \alpha \geq 0$ and $\beta + \alpha \geq 0$. Then

$$\begin{aligned}T_1(f+\beta) &= \vee_0^{f+\beta+\alpha} T - \alpha \vee_0^1 T \\ &= T_1(f+\beta+\alpha) - \alpha \vee_0^1 T\end{aligned}$$

while from facts $(b)\&(d)$, we infer that $T_1(f) = T_1(f+\beta+\alpha) - (\beta+\alpha) \vee_0^1 T$. Therefore,

$$T_1(f+\beta) = T_1(f) + T_1(\beta)$$

which proves that indeed T_1 is translation invariant. The same holds for $T_2 = T - T_1$.

As concerns the second part of Lemma 4.4.2, it suffices to prove that T_1 is upper continuous when T has this property.

Let $(f_n)_n$ be a decreasing sequence in $C(K)$, which converges pointwise to a function f also in $C(K)$. By Dini's lemma, $\|f_n - f\| \to 0$. Since T_1 is a Lipschitz operator (see Lemma 4.3.4), we conclude that $\|T_1(f_n) - T(f)\| \to 0$.

This ends the proof of Lemma 4.4.2. ■

Now it is clear that the representation Theorem 4.3.6 can be extended to the framework of comonotonic additive operators $T : C(K) \to E$ with bounded variation by considering Choquet–Bochner integrals associated with differences of vector capacities. The details are very close to the construction of integrals associated with the games with the bounded variation. See the Comments Sect. 2.10 at the end of Chap. 2.

4.5 The Volterra–Choquet-Type Operators

The Volterra operator,

$$V : L^2[0,1] \to L^2[0,1], \quad V(f)(x) = \int_0^x f(t)\,\mathrm{d}t, \; x \in [0,1], \tag{4.13}$$

is a classical example of compact linear operator whose spectrum reduces to $\{0\}$. Its theory is presented by Gohberg and Krein in their book [89]. In what follows we will discuss the generalized *Volterra-Choquet operator*

$$V : E \to E, \quad V(f)(x) = (C)\int_0^x f\,\mathrm{d}\mu, \; x \in [0,1], \tag{4.14}$$

obtained by replacing the space $L^2[0,1]$ by one of the spaces $C([0,1])$, or $L^p(\mu)$ (with $p \in [1,\infty)$), where μ is a finite capacity defined on the σ-algebra of Lebesgue measurable subsets of $[0,1]$ and playing some special properties such as lower and upper continuity, strong subadditivity, and completeness. For example, μ can be any distorted measure of the form $\mu(A) = \gamma(m(A))$, where m is the Lebesgue measure restricted to $[0,1]$ and $\gamma : [0,1] \to \mathbb{R}_+$ is any increasing, concave, and continuous function such that $\gamma(0) = 0$.

According to our discussion in Sect. 3.5, under these assumptions, all the capacitary spaces $L^p(\mu)$ with $p \in [1,\infty)$ are Banach lattices. They all contain the unity 1 and the same is true for the Banach lattice $C([0,1])$.

Theorem 4.5.1 *Let E be one of the Banach lattices $C([0,1])$ and $L^p(\mu)$ with $p \in [1,\infty)$. Then the formula*

$$V : E \to E, \quad V(f)(x) = (C)\int_0^x f\,\mathrm{d}\mu, \; x \in [0,1],$$

defines a continuous, weakly nonlinear, and comonotonic additive operator.

To prove this result, it suffices to show that V maps each of the spaces E into itself as a continuous map. All other properties (invariance to translation, comonotonic additivity and sublinearity) are consequences of the general theory of Choquet integral exposed in Sects. 2.4–2.6. The necessary details are provided by the following lemmas.

Lemma 4.5.2 *If $f \in C[0,1]$, then $V(f) \in C[0,1]$ and for all $f, g \in C[0,1]$, we have*

$$\|V(f) - V(g)\|_{C[0,1]} \leq \mu([0,1]) \cdot \|f - g\|_{C[0,1]}.$$

In particular, the norm of this variant of the Choquet–Volterra operator is $\mu([0,1])$.

Proof Suppose first that $f \geq 0$, and let $x_n, x \in [0,1]$ $(n \in \mathbb{N})$ with $x_n \to x$. We will prove that $V(f)(x_n) \to V(f)(x)$. For this, denote the sup norm of f by M and put

$$A_n(x,\alpha) = \{s \in [0,x_n] : f(s) \geq \alpha\} \text{ and } A(x,\alpha) = \{s \in [0,x] : f(s) \geq \alpha\}.$$

Then, according to the definition of the Choquet integral,

$$\begin{aligned}
&|V(f)(x_n) - V(f)(x)| \\
&= \left| \int_0^M \mu(\{s \in [0,x_n] : f(s) \geq \alpha\})\, \mathrm{d}\alpha - \int_0^M \mu(\{s \in [0,x] : f(s) \geq \alpha\})\, \mathrm{d}\alpha \right| \\
&\qquad = \left| \int_0^M \mu(A_n(x,\alpha))\, \mathrm{d}\alpha - \int_0^M \mu(A(x,\alpha))\, \mathrm{d}\alpha \right|.
\end{aligned}$$

Since a sequence $(y_n)_n$ is converging to a limit y if and only if

$$y = \liminf_{n\to\infty} y_n = \sup_{n\geq 1}\left(\inf_{k\geq n} y_k\right) \text{ and } y = \limsup_{n\to\infty} y_n = \inf_{n\geq 1}\left(\sup_{k\geq n} y_k\right),$$

in order to prove that $V(f)(x_n) \to V(f)(x)$, it suffices to consider the two cases when $x_n \nearrow x$ and $x_n \searrow x$.

If $x_n \nearrow x$, then $A_n(x,\alpha) \subset A_{n+1}(x,\alpha) \subset A(x,\alpha) \subset [0,1]$, for all n in $\mathbb{N}$ and $A(x,\alpha) = \bigcup_n A_n(x,\alpha)$. Since μ is lower continuous, it follows that $\lim_{n\to\infty} \mu(A_n(x,\alpha)) = \mu(A(x,\alpha))$ for all $\alpha \in [0,M]$, which in turn implies that $\lim_{n\to\infty} V(f)(x_n) = V(f)(x)$ according to the Monotone Convergence Theorem for the Lebesgue integral. The case when $x_n \searrow x$ can be treated in a similar way (by taking into account the upper continuity of μ). Therefore, $V(f) \in C[0,1]$ when $f \in C[0,1]$ and $f \geq 0$.

Suppose now that f is real-valued. Then $-M \leq f(t) \leq M$ for all $t \in [0,1]$ and some positive constant M. According to the definition of Choquet integral, we have

$$
\begin{aligned}
V(f)(x) = \int_0^M & \mu(\{t \in [0, x] : f(t) \geq \alpha\})\, \mathrm{d}\alpha \\
& + \int_{-M}^0 \mu(\{t \in [0, x] : f(t) \geq \alpha\})\, \mathrm{d}\alpha - M \cdot \mu([0, x]) \\
& = \int_0^M \mu(\{t \in [0, x] : f^+(t) \geq \alpha\})\, \mathrm{d}\alpha \\
& \quad + \int_0^M \mu(\{t \in [0, x] : f^-(t) \leq \alpha\})\, \mathrm{d}\alpha - M \cdot \mu([0, x])
\end{aligned}
$$

and the continuity of $V(f)$ follows from the above reasoning.

The Lipschitzianity of the operator V is a consequence of Theorem 3.1.1. Indeed, every continuous function on $[0, 1]$ is measurable and bounded and according to this theorem

$$
\begin{aligned}
|V(f)(x) - V(g)(x)| &= \left| (C) \int_0^x f \, \mathrm{d}\mu - (C) \int_0^x g \, \mathrm{d}\mu \right| \\
&\leq (C) \int_0^x |f - g| \, \mathrm{d}\mu \leq \mu([0, 1]) \cdot \|f - g\|_{C([0,1])} \, .
\end{aligned}
$$

Passing to the supremum in the left-hand side over $x \in [0, 1]$,, we infer that

$$
\|V(f) - V(g)\|_{C([0,1])} \leq \mu([0, 1]) \cdot \|f - g\|_{C([0,1])} \, .
$$

In particular, $\|V(f)\|_{C([0,1])} \leq \mu([0, 1]) \cdot \|f\|_{C([0,1])}$ for all f. Since this formula becomes an equality for $f = 1$, we conclude that the norm of V as an operator from $C([0, 1])$ into itself is $\mu([0, 1])$. ■

Lemma 4.5.3 *If $p \in [1, \infty)$ and f belongs to $L^p(\mu)$, then $V(f) \in L^p(\mu)$ and for all $f, g \in L^p(\mu)$ we have*

$$
\|V(f) - V(g)\|_{L^p(\mu)} \leq \mu([0, 1]) \cdot \|f - g\|_{L^p(\mu)}.
$$

In particular, $\|V\| \leq \mu([0, 1])$.

This result does not use the upper continuity of μ.

Proof Since μ is a finite capacity, all the spaces $L^p(\mu)$ are included into $L^1(\mu)$, the inclusions being continuous. See Remark 3.5.12. Notice also that $V(f)$ is a Lebesgue measurable function, whenever $f \in L^p(\mu)$. Indeed, $V(f)(x)$ equals

$$
\begin{aligned}
V(f)(x) = \int_0^{+\infty} & \mu(\{t \in [0,x] : f(t) \geq \alpha\})\, \mathrm{d}\alpha \\
& + \int_{-\infty}^{0} [\mu(\{t \in [0,x] : f(t) \geq \alpha\}) - \mu([0,x])]\, \mathrm{d}\alpha \\
& = \int_0^{\infty} \mu(\{t \in [0,x] : f^+(t) \geq \alpha\})\, \mathrm{d}\alpha \\
& + \lim_{n\to\infty} \int_n^0 [\mu(\{t \in [0,x] : f(t) \geq \alpha\}) - \mu([0,x])]\, \mathrm{d}\alpha \\
& = \int_0^{\infty} \mu(\{t \in [0,x] : f^+(t) \geq \alpha\})\, \mathrm{d}\alpha \\
& + \lim_{n\to\infty} \left[\int_0^n \mu(\{t \in [0,x] : f^-(t) \leq \alpha\})\, \mathrm{d}\alpha - n\mu([0,x]) \right],
\end{aligned}
$$

so $V(f)$ is the sum of an increasing function and the limit of a sequence of measurable functions (each one being the difference of two increasing functions).

Moreover, taking into account the modulus inequality, we infer that

$$
\begin{aligned}
(C)\int_0^1 |V(f)(x)|\, \mathrm{d}\mu(x) &= (C)\int_0^1 \left| (C)\int_0^x f(t)\, \mathrm{d}\mu(t) \right| \mathrm{d}\mu(x) \\
&\leq (C)\int_0^1 \left((C)\int_0^1 |f(t)|\, \mathrm{d}\mu(t) \right) \mathrm{d}\mu(x) \\
&= \mu([0,1]) \cdot \|f\|_{L^1(\mu)} < \infty
\end{aligned}
$$

and

$$
\begin{aligned}
(C)\int_0^1 |V(f)(x)|^p\, \mathrm{d}\mu(x) &= (C)\int_0^1 \left| (C)\int_0^x f(t)\, \mathrm{d}\mu(t) \right|^p \mathrm{d}\mu(x) \\
&\leq (C)\int_0^1 \left((C)\int_0^1 |f(t)|\, \mathrm{d}\mu(t) \right)^p \mathrm{d}\mu(x) \\
\leq (C)\int_0^1 & \left[\left((C)\int_0^1 |f(t)|^p\, \mathrm{d}\mu(t) \right) \left((C)\int_0^1 1\, \mathrm{d}\mu(t) \right)^{p/q} \right] \mathrm{d}\mu(x) \\
&\leq \mu([0,1])^p \cdot \|f\|^p_{L^p(\mu)} < \infty
\end{aligned}
$$

if $p \in (1,\infty)$ and $1/p + 1/q = 1$. Therefore, V maps the space $L^p(\mu)$ into itself and, according to Lemma 4.1.12, it is a Lipschitz operator of norm at most $\mu([0,1])$. ■

Open Problem 4.5.4 *In the classical case of Volterra operator* 4.13, *it is known that its norm equals* $2/\pi$, *so the estimates provided by Lemma* 4.5.3 *are not the best. The problem of computing the exact values of* $\|V\|$ *is left open.*

The classical Volterra operator has many nice properties. For example, it is *compact* (even of Hilbert-Schmidt type) and *cyclic* (i.e., there are vectors x in $L^2([0,1])$ whose linear span of their orbits $\{x, V(x), V^2(x), \ldots\}$ are dense in $L^2([0,1])$).

In the context of capacitary spaces, the compactness holds only for the Volterra-Choquet type operators acting on the positive cones $L_+^p(\mu)$ with $p \in (1,\infty)$ or on the whole space $C([0,1])$. The main ingredient in the proof is the same as in the classical case, the Arzelá–Ascoli criterion of compactness:

Theorem 4.5.5 (The Arzelá-Ascoli Theorem) *Suppose that* K *is a compact metric space with metric* d. *Then the closure of a subset* $\mathcal{F}$ *of* $C(K)$ *is relatively compact if and only if it verifies the following two conditions:*

(i) (uniform boundedness) $\sup\{\|f\| : f \in \mathcal{F}\} < \infty$;
(ii) (equicontinuity) *for every* $\varepsilon > 0$, *there exists* $\delta > 0$ *such that whenever* $x, y \in K$ *with* $d(x,y) < \delta$, *we have* $|f(x) - f(y)| < \varepsilon$ *for every* $f \in \mathcal{F}$.

For details, see Carothers [32], Theorem 11.18, p. 181.

The application of the Arzelá–Ascoli Theorem is made possible by the following result.

Lemma 4.5.6 *If* $p \in (1,\infty)$, *then the Volterra–Choquet-type operator*

$$V : f \to V(f), \quad V(f)(x) = (C)\int_0^x f \, d\mu,$$

verifies the inequality

$$|V(f)(x) - V(f)(y)| \leq \|f\|_{L_+^p(\mu)} \cdot \mu([x,y])^{1-1/p}$$

for all $f \in L_+^p(\mu)$ *and* $0 \leq x \leq y \leq 1$, *and thus, it maps* $L_+^p(\mu)$ *into* $C_+([0,1])$.

Proof Since μ is strongly subadditive, it follows that the associated Choquet integral is a subadditive functional. See Theorem 2.6.1. Taking into account that $[0,y] = [0,x] \cup [x,y]$, it follows that

$$V(f)(y) \leq V(f)(x) + (C)\int_x^y f \, d\mu,$$

which, by Hölder's inequality, implies that

$$0 \le V(f)(y) - V(f)(x) \le (C) \int_x^y f \, d\mu \le \|f\|_{L^p(\mu)} \cdot \mu([x, y])^{1-1/p}.$$

■

Theorem 4.5.7 *The Volterra–Choquet-type operator* $V : f \to V(f)$*, defined by* $V(f)(x) = (C) \int_0^x f \, d\mu$*, defines a weakly nonlinear compact operator from* $L_+^p(\mu)$ *into* $C_+([0, 1])$ *(and, a fortiori, from* $L_+^p(\mu)$ *into itself) for every* p *in the interval* $(1, \infty)$.

Proof The fact that V maps $L_+^p(\mu)$ into $C_+([0, 1])$ follows from Lemma 4.5.6, which also implies that V maps the bounded subsets of $L_+^p(\mu)$ into bounded and equicontinuous subsets of $C_+([0, 1])$. The proof ends by reminding the fact that all bounded and equicontinuous subsets of $C([0, 1])$ are relatively compact, according to the Arzelá–Ascoli Theorem. Since the inclusion $C_+([0, 1]) \hookrightarrow L_+^p(\mu)$ is continuous, it maps the relatively compact sets into relatively compact sets. Consequently, V defines also a compact operator from $L_+^p(\mu)$ into itself. ■

An inspection of the argument of Lemma 4.5.6, shows that

$$|V(f)(x) - V(f)(y)| \le \|f\|_{C([0,1])} \cdot \mu([x, y]) \quad \text{for all } 0 \le x \le y \le 1,$$

so that, proceeding as in the proof of Theorem 4.5.7, we infer the following complementary result:

Theorem 4.5.8 *The Volterra–Choquet operator* $V : C[0, 1] \to C[0, 1]$ *defined by the formula*

$$V(f)(x) = (C) \int_0^x f \, d\mu \quad \textit{for } f \in C([0, 1]) \textit{ and } x \in [0, 1],$$

is a weakly nonlinear monotone and compact operator.

Open Problem 4.5.9 *In the case of a* σ*-additive measure* μ *defined on* $\mathcal{B}([0, 1])$*, the Volterra-type operator acting on* $L^p(\mu)$ *is compact for every* $p \in (1, \infty)$ *(, and this fact can be deduced from Theorem* 4.5.7*). In the framework of capacities (lower continuous and strongly subadditive), we were able to prove the property of compactness only for the restriction of these operators to the positive cone* $L_+^p(\mu)$*. It is open whether this result is the best possible or not.*

4.6 Hölder's Inequality for the Monotone and Sublinear Operators

The classical inequality of Hölder was extended in Sect. 3.1 to the framework of Choquet integrals. A similar argument, based on Young's inequality, allows us to extend it to the more general framework of sublinear and monotone operators. For the convenience of the reader, we recall here Young's inequality: if $p, q \in (1, \infty)$ and $1/p + 1/q = 1,$, then

$$uv \leq \frac{u^p}{p} + \frac{v^q}{q} \quad \text{for all } u, v \geq 0. \tag{4.15}$$

Equality occurs if and only if $u^p = v^q$.

Theorem 4.6.1 (Hölder's Inequality for $p, q \in (1, \infty)$ and $1/p + 1/q = 1$)) *Suppose that X and Y are two Hausdorff topological spaces and E and F are respectively vector sublattices of $C_b(X)$ and $C_b(Y)$, which contain the constant function 1. Then every sublinear and monotone operator $T : E \to F$ such that $T(1) = 1$ verifies the inequality*

$$T(|fg|) \leq [T(|f|^p)]^{1/p} \cdot [T(|g|^q)]^{1/q}. \tag{4.16}$$

for all $f, g \in E$ such that $fg \in E$.

Proof For $y \in Y$ arbitrarily fixed, consider the sublinear and monotone functional $A_y : E \to \mathbb{R}$ defined by the formula

$$A_y(f) = (T(f))(y).$$

Clearly, $A_y(1) = 1$.

Assuming $A_y(|f|^p) > 0$ and $A_y(|g|^q) > 0,$, we apply the inequality (4.15) for $u = |f|/A_y(|f|^p)^{1/p}$ and $v = |g|/A_y(|g|^q)^{1/q}$ to infer that

$$\frac{|f|}{A_y(|f|^p)^{1/p}} \frac{|g|}{A_y(|g|^q)^{1/q}} \leq \frac{1}{p} \cdot \frac{|f|^p}{A_y(|f|^p)} + \frac{1}{q} \cdot \frac{|g|^q}{A_y(|g|^q)}. \tag{4.17}$$

Since the functional A_y is monotone and sublinear, the last inequality implies

$$\frac{A_y(|fg|)}{A_y(|f|^p)^{1/p} \cdot A_y(|g|^q)^{1/q}} \leq \frac{1}{p} + \frac{1}{q} = 1,$$

that is, $T(|f \cdot g|)(y) \leq [T(|f|^p)(y)]^{1/p} \cdot [[T(|g|^q)(y)]^{1/q}$, which is inequality (4.16) in the statement.

If $A_y(|f|^p) = 0$ and/or $A_y(|g|^q) = 0$,, then one has to repeat the above reasoning and to replace in (4.17) the vanishing number(s) by an $\varepsilon > 0$ arbitrarily small. Finally, by passing to the limit as $\varepsilon \to 0$, we conclude that $A_y(|f| \cdot |g|) = 0$. The proof is done. ∎

Applications of Theorem 4.6.1 will be presented in Sects. 5.1 and 5.2.

If an operator $T : E \to F$ is monotone and subadditive, then

$$|T(f)| \leq T(|f|) \quad \text{for all } f \in E, \tag{4.18}$$

which leads to *Hölder's inequality for $p = 1$ and $q = \infty$*:

$$|T(fg)| \leq T(|fg|) \leq T(|f|) \sup_{x \in X} |g(x)| \tag{4.19}$$

for all $f, g \in E$ such that $fg \in E$.

If X is a locally compact Hausdorff space and $T : C_b(X) \to \mathbb{R}$ is a positive linear functional for which $T(1) = 1$, then T admits the integral representation $T(f) = \int_X f d\mu$ for a suitable Borel probability measure μ and the difference

$$T(f^2) - T(f)^2 = \int_X f^2 \mathrm{d}\mu - \left(\int_X f \mathrm{d}\mu \right)^2$$

is just the *variance* of f. The fact that the variance is nonnegative follows from the Cauchy–Bunyakovsky–Schwarz inequality (the particular case of Hölder's inequality for $p = q = 2$). Thus, in the general context of sublinear and monotone operators $T : C_b(X) \to C_b(X)$, the quantity

$$D_T^2(f) = T(1) \cdot T(f^2) - T(f)^2$$

can be interpreted as the *T-variance* of f. The *T-covariance* of a pair of functions f and g in $C_b(X)$ can be introduced via the formula

$$Cov_T(f, g) = T(1) \cdot T(fg) - T(f)T(g).$$

Open Problem 4.6.2 *Under what conditions on T the following nonlinear version of the Cauchy–Bunyakovsky–Schwarz inequality,*

$$|Cov_T(f, g)| \leq \sqrt{D_T^2(f)} \sqrt{D_T^2(g)},$$

is true?

Some results related to this open problem are presented in what follows. We need the following auxiliary result

Lemma 4.6.3 *If T is a monotone increasing and sublinear operator that maps $C_b(X)$ into itself, then*

$$D_T^2(-|f|) = T(1) \cdot T(|f|^2) - |T(-|f|)|^2 \geq 0,$$

for all $f \in C_b(X)$.

Proof Since T is monotone and subadditive, the fact that $0 \leq (\lambda - |f(x)|)^2$ for all $\lambda > 0$ and $x \in X$ yields

$$0 \leq T[(\lambda - |f|)^2](x) \leq \lambda^2 T(1)(x) + 2\lambda T(-|f|)(x) + T(|f|^2)(x). \tag{4.20}$$

Suppose by reduction ad absurdum that there exists $x_0 \in X$ such that

$$|T(-|f|)(x_0)| > \sqrt{T(1)(x_0) \cdot T(f^2)(x_0)}. \tag{4.21}$$

Then the second degree polynomial in λ,

$$\lambda^2 T(1)(x_0) + 2\lambda T(-|f|)(x_0) + T(|f|^2)(x_0),$$

will have two positive distinct solutions $\lambda_1 < \lambda_2$. As a consequence, for any $\lambda \in (\lambda_1, \lambda_2)$,

$$\lambda^2 T(1)(x_0) + 2\lambda T(-|f| \cdot |g|)(x_0) + T(f^2)(x_0) < 0,$$

in contradiction with condition (4.20). Therefore (4.21) does not hold and the proof of the lemma is done. ■

The next lemma provides a partial answer to Problem 4.6.2.

Lemma 4.6.4 *Suppose that $T : C_b(X) \to C_b(X)$ is a Choquet-type operator. Then for all pairs of functions $f, g \in C_b(X)$ such that $|f|$ and $|g|$ are comonotonic, we have the inequality*

$$|Cov_T(-|f|, -|g|)| \leq \sqrt{D_T^2(-|f|)}\sqrt{D_T^2(-|g|)}.$$

Proof Let $\lambda > 0$ arbitrarily fixed. According to Lemma 4.6.3,

$$|T(-|f| - \lambda|g|)|^2 \leq T(1) \cdot T(|f|^2 + 2\lambda|fg| + \lambda^2|g|^2),$$

while the fact that T is comonotonic additive yields

$$|T(-|f| - \lambda |g|)|^2 = (T(-|f|) + \lambda T(-|g|))^2 .$$

Therefore,

$$\lambda^2 D^2(-|g|) + 2\lambda (T(1) \cdot T(|fg|) - T(-|f|)T(-|g|)) + D^2(-|f|) \geq 0$$

and taking into account that $\lambda > 0$ was arbitrarily fixed one can conclude (repeating the argument used in the proof of Lemma 4.6.3) that

$$|T(1) \cdot T(|fg|) - T(-|f|)T(-|g|)|^2 \leq D_T^2(|f|)D_T^2(|g|),$$

which ends the proof. ■

4.7 Comments

The sublinear operators play an important role in functional analysis and convex analysis, a well-known example being the Hahn–Banach extension theorem (and its numerous consequences). Among the hundreds of works on this subject, we mention here those by Feldman [62], Kantorovich [105], Kusraev and Kutateladze [110], Niculescu and Persson [137], and Rubinov [158, 159].

The notion of weakly nonlinear operator appeared in a series of papers by Gal and Niculescu [75–79, 81], devoted to the extension of Korovkin's theorem to a nonlinear framework.

Theorem 4.1.10, concerning the continuity of weakly nonlinear and monotone operators, is a generalization of a classical result of M. G. Krein [109] concerning the continuity of positive linear functionals. In this form it appears here for the first time. See [1] and [136] for related results.

The material presented in Sects. 4.2–4.4 is based on the paper [76].

Volterra operator has been studied and continues to be studied by many authors. Two important books on Volterra's operator are those by Brodskii [28] and Gohberg-Krein [89]. Halmos' Problem book [96] also contains several nice results related to Volterra's operator. The asymptotic behavior of the norm $\|V^n\|$ is described in Little-Reade [118]. An interesting fact about the Volterra operator is the determination of its invariant subspace lattice (see Conway [44], Chapter 4 and Brodskii [27], Dixmier [53], Donoghue [56], Kalisch [103], and Sarason [161]). The properties of compactness and cyclicity were studied by Gallardo-Montes [83], León-Saavedra–Piqueras-Lerena [114, 115], and Salas [160].

The presentation of Volterra-type operators in the framework of Choquet integral follows the paper of Gal [71].

Let us mention that the study of Volterra type operators is just at the beginning and the status of many classical results remains open in this generalized framework. In particular, so is the problem of cyclicity. A positive partial result (included in [71]) is as follows.

Theorem 4.7.1 *Consider the Volterra-type operator*

$$V(f)(x) = (\mathrm{C}) \int_0^x f \,\mathrm{d}m_\varphi, \quad f \in C([0,1])$$

associated with the distorted Lebesgue measure

$$m_\varphi(A) = \varphi(m(A)),$$

where $\varphi : [0,1] \to \mathbb{R}$ *is the concave function defined by the formula* $\varphi(x) = 1 - e^{-x}$. *Then the unity* 1 *is a cyclic vector for both operators* V *and* $\mathrm{I} + V$.

In [69], Gal started the investigation of an analogue of the classical Volterra integral equation of the second kind, in which the Lebesgue integral is replaced by the Choquet integral associated with a capacity defined on the σ-algebra $\mathcal{B}([a,b])$ of Borel subsets of the interval $[a,b]$. Precisely, he focused on the integral equation of the form

$$\varphi(x) = f(x) + \lambda \cdot (\mathrm{C}) \int_a^x K(x,s,\varphi(s))\,\mathrm{d}\mu(s), \quad x \in [a,b], \tag{4.22}$$

with the given data $\lambda \in \mathbb{R}$, $f : [a,b] \to \mathbb{R}$, $K : [a,b] \times [a,b] \times \mathbb{R} \to \mathbb{R}$ and the unknown function $\varphi : [a,b] \to \mathbb{R}$.

It is clear that due to the generality of the Choquet integral, one cannot expect that such a nonlinear integral equation has solutions for any choice of the capacity μ. However, Gal was able to prove the following result.

Theorem 4.7.2 *Consider a kernel* $K \in C([a,b] \times [a,b] \times \mathbb{R})$, $K = K(x,s,u)$ *which is Lipschitz with respect to the third variable, i.e., for some constant* $L > 0$ *it verifies the inequality*

$$|K(x,s,u) - K(x,s,v)| \le L \cdot |u - v| \quad \textit{for all } x,s \in [a,b] \textit{ and } u,v \in \mathbb{R}.$$

Consider also a strongly subadditive and lower continuous capacity $\mu : \mathcal{B}([a,b]) \to [0,+\infty)$ *for which there exists a suitable function* $c : (0,\infty) \to (0,\infty)$ *such that* $\lim_{\tau\to\infty} c(\tau) = 0$ *and*

$$(\mathrm{C}) \int_a^x e^{\tau(s-x)}\,\mathrm{d}\mu(s) \le c(\tau) \quad \textit{for all } x \in [a,b] \textit{ and } \tau > 0,$$

Then for every $f \in C([a,b])$, the Volterra-Choquet integral equation (4.22) *has a unique solution $\varphi \in C([a,b])$.*

One can find easily examples of distorted capacities, which verify the hypotheses of Theorem 4.7.2. So is the case of $\mu = \sqrt{m}$, where m denotes the restriction of the Lebesgue measure to $[a,b]$.

The extension of Hölder's inequality presented in Sect. 4.6 (together with the equality case) is taken from Gal and Niculescu [77].

A discussion about the iterates of monotone increasing sublinear operators, illustrated with examples from Choquet integral theory, can be found in the article of Gal [72].

It is important noticing the existence of a notion of strong subadditivity in the framework of functionals defined on ordered Banach spaces E or just on their positive cones E_+. Precisely, denoting by $\mathcal{C}$ one of these alternatives, we say that $\Phi : \mathcal{C} \to \mathbb{R}$ is *strongly subadditive* functional if

$$\Phi(0) \geq 0 \text{ and } \Phi(\mathbf{x}+\mathbf{y}+\mathbf{z}) + \Phi(\mathbf{z}) \leq \Phi(\mathbf{x}+\mathbf{z}) + \Phi(\mathbf{y}+\mathbf{z})$$

for all $\mathbf{x}, \mathbf{y}, \mathbf{z} \in \mathcal{C}$; Φ is called *strongly superadditive* if $-\Phi$ is strongly subadditive.

Notice that every strongly additive functional Φ is also subadditive, i.e.,

$$\Phi(\mathbf{x}+\mathbf{y}) \leq \Phi(\mathbf{x}) + \Phi(\mathbf{y})$$

for all $\mathbf{x}, \mathbf{y} \in \mathcal{C}$ and every subadditive functional verifies the condition $\Phi(0) \geq 0$.

As noticed in [135], a continuous function $\Phi : \mathbb{R}_+ \to \mathbb{R}$ is strongly subadditive if and only if Φ is concave and $\Phi(0) \geq 0$ (respectively Φ is strongly superadditive if and only if Φ is convex and $\Phi(0) \leq 0$). As a consequence, the *Shannon entropy*,

$$\Phi(\mathbf{x}) = -\sum\nolimits_{k=1}^{N} x_k \log x_k, \quad \mathbf{x} \in \mathbb{R}_+^N,$$

is strongly subadditive, while the square of the Euclidean norm is strongly superadditive on the same cone $\mathbf{x} \in \mathbb{R}_+^N$.

The property of strong subadditivity can be characterized via the *differences of second order,*

$$\Delta_{\mathbf{x}}\Delta_{\mathbf{y}}\Phi(\mathbf{z}) = \Phi(\mathbf{x}+\mathbf{y}+\mathbf{z}) - \Phi(\mathbf{x}+\mathbf{z}) - \Phi(\mathbf{y}+\mathbf{z}) + \Phi(\mathbf{z})$$

which leads to several useful criteria of strong subadditivity using differential calculus.

Theorem 4.7.3 (See [135], Theorem 3) *Suppose that E is a finite-dimensional ordered Banach space and $\Phi : E_+ \to F$ is a continuous function, differentiable on E_{++}. Then*

Φ is strongly superadditive if and only if $\Phi(\mathbf{0}) \leq 0$ and the differential of Φ is monotone increasing as a map from E_+ into $L(E, F)$.

Theorem 4.7.4 (See [135], Theorem 4) *Suppose that Φ belongs either to $C(\mathbf{R}_+^N) \cap C^2(\mathbb{R}_{++}^N)$ or to $C^2(\mathbb{R}^N)$. Then Φ is strongly superadditive if and only if $\Phi(\mathbf{0}) \leq 0$ and all partial derivatives of second-order $\frac{\partial^2 \Phi}{\partial x_i \partial x_j}$ are nonnegative.*

As a consequence, it is recovered the known fact that the monotone increasing functionals $A \to \det A$ and $A \to \operatorname{trace} A^p$ $(1 \leq p \leq 2)$ are superadditive on the cone $\mathrm{Sym}^+(N, \mathbb{R})$, of all positive matrices.

According to Theorem 2.6.1, a condition equivalent to strong subadditivity of a capacity is the submodularity of the Choquet functional associated with it. A functional Φ defined on an ordered Banach space E, or only its positive cone, is said to be *submodular* if

$$\Phi(\mathbf{x} \vee \mathbf{y}) + \Phi(\mathbf{x} \wedge \mathbf{y}) \leq \Phi(\mathbf{x}) + \Phi(\mathbf{y}) \quad \text{for all } \mathbf{x}, \mathbf{y};$$

Φ is called *supermodular* if $-\Phi$ is submodular. A straightforward characterization of these functional in the case of the Euclidean space endowed with the coordinatewise order is as follows:

Theorem 4.7.5 (Topkis' Characterization Theorem; See [130]) *Suppose that $\mathcal{C}$ is either the Euclidean space $\mathbb{R}^N$ or its positive cone and $\Phi : \mathcal{C} \to \mathbb{R}$ is a functional continuous on $\mathcal{C}$ and continuously differentiable on $\operatorname{int} \mathcal{C}$. Then Φ is supermodular if and only if*

$$\frac{\partial^2 \Phi}{\partial x_i \partial x_j}(\mathbf{x}) \geq 0 \quad \text{for all } \mathbf{x} \in \operatorname{int} \mathcal{C} \text{ and } i \neq j.$$

Combining this result with Theorem 4.7.4, we arrive at the following consequence.

Corollary 4.7.6 *Every strongly subadditive (respectively strongly superadditive) functional that belongs either to $C(\mathbf{R}_+^N) \cap C^2(\mathbb{R}_{++}^N)$ or to $C^2(\mathbb{R}^N)$, is submodular (respectively supermodular).*

The converse does not work in general. See the case of the log-sum-exp function, defined on $\mathbb{R}^N$ by the formula

$$\mathrm{LSE}(\mathbf{x}) = \log \left(\frac{1}{N} \sum_{k=1}^{N} e^{x_k} \right), \quad \mathbf{x} \in \mathbb{R}^N.$$

Clearly,

$$\frac{\partial^2 \mathrm{LSE}}{\partial x_i \partial x_j}(\mathbf{x}) = -\frac{e^{x_i} e^{x_j}}{\left(\sum_{k=1}^{N} e^{x_k}\right)^2} < 0 \quad \text{for all } i \neq j,$$

so by Topkis' characterization theorem, the function LSE is submodular. Moreover, $\mathrm{LSE}(\mathbf{0}) = 0$ but this function is not strongly subadditive. For example, for $N = 2$ and $x = (1, 0)$, $y = (0, 1)$ and $z = (1, 1)$, we have

$$\begin{aligned} &\mathrm{LSE}(\mathbf{x} + \mathbf{y} + \mathbf{z}) + \mathrm{LSE}(\mathbf{z}) - \mathrm{LSE}(\mathbf{x} + \mathbf{z}) - \mathrm{LSE}(\mathbf{y} + \mathbf{z}) \\ &\qquad = \log\left(\frac{1}{2}\left(e^2 + e^3\right)\right) + \log\left(\frac{1}{2}(e + e)\right) \\ &\qquad\qquad - \log\left(\frac{1}{2}\left(e^2 + e\right)\right) - \log\left(\frac{1}{2}\left(e + e^2\right)\right) > 0.379\,. \end{aligned}$$

Somehow surprisingly, the log-sum-exp function verifies the following property of comonotonic strong superadditivity:

$$\mathrm{LSE}(\mathbf{x} + \mathbf{y} + \mathbf{z}) + \mathrm{LSE}(\mathbf{z}) \geq \mathrm{LSE}(\mathbf{x} + \mathbf{z}) + \mathrm{LSE}(\mathbf{y} + \mathbf{z})$$

for all $\mathbf{x}, \mathbf{y}, \mathbf{z} \in \mathbb{R}_+^N$ such that $\mathbf{x}$ and $\mathbf{y}$ are comonotonic. This is a direct consequence of Chebyshev's algebraic inequality,

$$\langle \mathbf{u}, \mathbf{p} \rangle \langle \mathbf{v}, \mathbf{p} \rangle \leq \langle \mathbf{uv}, \mathbf{p} \rangle$$

whenever $\mathbf{u}, \mathbf{v} \in \mathbb{R}^N$ are comonotonic and $\mathbf{p} = (p_1, \ldots, p_N)$ is a probability vector (that is, $p_k \geq 0$ for all k and $\sum_{k=1}^{N} p_k = 1$). See [137], Section 2.2, Exercise 8.

Korovkin's Theorems in the Framework of Weakly Nonlinear Operators 5

One of the most elegant results in approximation theory is Korovkin's theorem, inspired by Bernstein's approach to Weierstrass's famous approximation theorem.

Theorem (Korovkin's Theorem [107, 108]) *Let $(L_n)_n$ be a sequence of positive linear operators that map $C([0, 1])$ into itself. Suppose that the sequence $(L_n(f))_n$ converges to f uniformly, for each of the following three test functions:*

$$e_k : x \to x^k, \ k = 0, 1, 2.$$

Then the sequence $(L_n(f))_n$ converges to f uniformly for all functions f in $C([0, 1])$.

A similar result holds if $C([0, 1])$ is replaced by the space $C_{2\pi}(\mathbb{R})$ (of continuous and periodic functions of period 2π) and the triplet of test functions is replaced by the functions 1, $\cos$, and $\sin$.

The sequence of Bernstein operators $B_n : C([0, 1]) \to C([0, 1])$, defined by

$$B_n(f)(x) = \sum_{k=0}^{n} f\left(\frac{k}{n}\right)\binom{n}{k}x^k(1-x)^{n-k}, \quad \text{for } x \in [0, 1] \text{ and } n \in \mathbb{N},$$

verifies the conditions of Korovkin's Theorem since

$$B_n(e_0)(x) = 1, \ B_n(e_1)(x) = x \text{ and } B_n(e_2)(x) = \frac{n-1}{n}x^2 + \frac{1}{n}x.$$

This is a direct consequence of the following combinatorial identities:

S. G. Gal, C. P. Niculescu, *Choquet Integral and Monotone Sublinear Operators*, Frontiers in Mathematics, https://doi.org/10.1007/978-3-031-71701-7_5

$$\sum_{k=0}^{n} \binom{n}{k} t^k (1-t)^{n-k} = 1$$

$$\sum_{k=0}^{n} k \binom{n}{k} t^k (1-t)^{n-k} = nt$$

$$\sum_{k=0}^{n} k^2 \binom{n}{k} t^k (1-t)^{n-k} = nt(1-t+nt)$$

whose proof can be found in the book of Choudary and Niculescu [43, Theorem 8.8.1, p. 256]. As a consequence, for every $f \in C([0, 1])$,

$$B_n(f) \to f \quad \text{uniformly on } [0, 1],$$

which is the famous approximation result due to Bernstein [23].

Over the years, many generalizations of Korovkin's Theorem appeared, in a variety of settings involving important Banach function spaces. A few of them are recalled in the next section, but a nice account on this direction of research is offered by the authoritative monograph of Altomare and Campiti [12] and the excellent survey of Altomare [7].

Motivated by the theory of Choquet integrability, the authors of this book published a series of papers extending Korovkin's results to the class of weakly nonlinear operators. See [75–79, 81]. This chapter is aimed to present these results as well as the very recent contributions by Altomare [8–11] and Popa [152, 153]. Many types of convergence are investigated, including the uniform convergence, the pointwise (almost everywhere) convergence, the convergence in measure, the L^p-convergence, etc. The theory is illustrated by concrete examples, most of them based on the theory of Choquet integral.

5.1 The Case of Operators Defined on Subspaces of $C_b(\Omega)$ with $\Omega \subset \mathbb{R}^N$

In the case of linear and positive operators acting on spaces of continuous functions defined on a compact subset of $\mathbb{R}^N$, the role of Korovkin's triplet $\{1, x, x^2\}$ of test functions is played by a family of $N+2$ functions, consisting of the function identically 1, the canonical projections,

$$\mathrm{pr}_k : \mathbb{R}^N \to \mathbb{R}, \qquad \mathrm{pr}_k(x_1, \ldots, x_N) = x_k, \ k = 1, \ldots, N$$

and the function $\sum_{k=1}^{N} \mathrm{pr}_k^2$. As was noticed in [75], this family needs to be enlarged when dealing with weakly nonlinear and monotone increasing operators.

Theorem 5.1.1 (The Nonlinear Version of Korovkin's Theorem on Locally Compact Sets in $\mathbb{R}^N$) *Suppose that Ω is a locally compact subset of the Euclidean space $\mathbb{R}^N$ and E is a vector sublattice of $\mathcal{F}(\Omega)$ that contains the test functions $1, \ \pm \mathrm{pr}_1, \ldots, \ \pm \mathrm{pr}_N$ and $\sum_{k=1}^N \mathrm{pr}_k^2$.*

(i) *If $(T_n)_n$ is a sequence of monotone increasing and sublinear operators from E into $\mathcal{F}(\Omega)$ such that*

$$\lim_{n\to\infty} T_n(f) = f \quad \text{uniformly on the compact subsets of } \Omega \tag{5.1}$$

for each of the $2N + 2$ aforementioned test functions, then the property (5.1) also holds for all nonnegative functions f in $E \cap C_b(\Omega)$.

(ii) *If, in addition, each operator T_n is translatable, then $(T_n(f))_n$ converges to f uniformly on the compact subsets of Ω, for every f in $E \cap C_b(\Omega)$.*

(iii) *In both cases* (i) *and* (ii), *the family of test functions can be reduced to $1, \ -\mathrm{pr}_1, \ldots, \ -\mathrm{pr}_N$ and $\sum_{k=1}^N \mathrm{pr}_k^2$ when Ω is included in the positive cone of $\mathbb{R}^N$.*

(iv) *If $f \in E$ is uniformly continuous and bounded on Ω and the operators T_n are weakly nonlinear and monotone increasing, then the convergence of $(T_n(f))_n$ to f is uniform on Ω provided that the property* of *convergence (5.1) holds uniformly on Ω for all test functions.*

Proof

(i) In order to make more easier the handling of the test functions, we denote

$$e_0 = 1, \ e_k = \mathrm{pr}_k \ \ (k = 1, \ldots, N) \text{ and } e_{N+1} = \sum_{k=1}^{N} \mathrm{pr}_k^2 .$$

Then, by replacing each operator T_n by $T_n/T_n(e_0)$, we may assume that $T_n(e_0) = 1$ for n sufficiently large.

Let $f \in E \cap C_b^+(\Omega)$ and let K be a compact subset of Ω. Then for every $\varepsilon > 0$, there is $\tilde{\delta} > 0$ such that

$$|f(\mathbf{x}) - f(\mathbf{y})| \le \varepsilon \quad \text{for every } \mathbf{y} \in K \text{ and } \mathbf{x} \in \Omega \text{ with } \|\mathbf{x} - \mathbf{y}\| \le \tilde{\delta};$$

this can be easily proved by reductio ad absurdum. If $\|\mathbf{x} - \mathbf{y}\| \ge \tilde{\delta}$, then

$$|f(\mathbf{x}) - f(\mathbf{y})| \le \frac{2\|f\|_\infty}{\tilde{\delta}^2} \cdot \|\mathbf{x} - \mathbf{y}\|^2,$$

so that, letting $\delta = 2\|f\|_\infty/\tilde{\delta}^2$, we obtain the estimate

$$|f(\mathbf{x}) - f(\mathbf{y})| \le \varepsilon + \delta\|\mathbf{x} - \mathbf{y}\|^2 \le \varepsilon + \delta\left(\sum_{k=1}^{N} x_k^2 - 2\sum_{k=1}^{N} x_k y_k + \|\mathbf{y}\|^2\right) \tag{5.2}$$

for all $\mathbf{y} \in K$ and $\mathbf{x} \in \Omega$. Since K is a compact set, it can be embedded into an N-dimensional cube $[a, b]^N$ for suitable $b \ge 0 \ge a$ and the estimate (5.2) yields

$$|f - f(\mathbf{y})e_0| \le \varepsilon e_0 + \delta\left[e_{N+1} + 2\sum_{k=1}^{N}(e_k(\mathbf{y}) - a)(-e_k) - 2a\sum_{k=1}^{N} e_k + \|\mathbf{y}\|^2 e_0\right].$$

Since the operators T_n are subadditive and positively homogeneous, they verify the inequality $|T_n(f) - T_n(g)| \le T_n(|f - g|)$ (see Remark 4.1.14) and thus,

$$|T_n(f) - f(\mathbf{y})| = |T_n(f) - T_n(f(\mathbf{y})e_0)| \le T_n(|f - f(\mathbf{y})e_0|) \le \varepsilon T_n(e_0) + \delta\left[T_n(e_{N+1}) + 2\sum_{k=1}^{N}(e_k(\mathbf{y}) - a)\,T_n(-e_k) - 2a\sum_{k=1}^{N} T_n(e_k) + \|\mathbf{y}\|^2\right]$$

for n sufficiently large and $\mathbf{y} \in K$. Here we used the assumption that f is nonnegative. By our hypothesis,

$$T_n(e_{N+1})(\mathbf{y}) + 2\sum_{k=1}^{N}(e_k(\mathbf{y}) - a)\,T_n(-e_k)(\mathbf{y}) - 2a\sum_{k=1}^{N} T_n(e_k)(\mathbf{y}) + \|\mathbf{y}\|^2 \to 0$$

uniformly on K as $n \to \infty$. Therefore,

$$\limsup_{n\to\infty}|T_n(f)(\mathbf{y}) - f(\mathbf{y})| \le \varepsilon,$$

whence we conclude that $T_n(f) \to f$ uniformly on K because ε was arbitrarily fixed.

(ii) Suppose in addition that each operator T_n is also translatable. Then $T_n(f + \|f\|e_0) = T_n(f) + \|f\| \cdot T_n(e_0)$. According to the assertion (i),

$$T_n(f + \|f\|e_0) \to f + \|f\|e_0, \quad \text{uniformly on } K.$$

Therefore, $T_n(f) \to f$ uniformly on K.

(iii) When K is included in the positive cone of $\mathbb{R}^N$, it can be embedded into an N-dimensional cube $[0, b]^N$ for a suitable $b > 0$ and the estimate (5.2) yields

$$|f - f(\mathbf{y})e_0| \le \varepsilon e_0 + \delta \left[e_{N+1} + 2\sum_{k=1}^{N} e_k(\mathbf{y})\,(-e_k) + \|\mathbf{y}\|^2\, e_0 \right].$$

Proceeding as above, we infer that

$$|T_n(f)(\mathbf{y}) - f(\mathbf{y})| \le \varepsilon T_n(e_0) + \delta \left[T_n(e_{N+1})(\mathbf{y}) + 2\sum_{k=1}^{N} e_k(\mathbf{y}) T_n(-e_k)(\mathbf{y}) + \|\mathbf{y}\|^2 \right]$$

for n sufficiently large and $\mathbf{y} \in K$, provided that $f \ge 0$. As a consequence, in both cases (i) and (ii), the family of testing functions can be reduced to $e_0, -e_1, \ldots, -e_N$ and e_{N+1}.

(iv) When dealing with functions $f \in E$ uniformly continuous and bounded on Ω, an inspection of the argument above shows that f verifies an estimate of the form (5.2) for all $\mathbf{x}, \mathbf{y} \in \Omega$, a fact that implies the convergence of $(T_n(f))_n$ to f uniformly on Ω. ∎

Clearly, Theorem 5.1.1 extends both Korovkin's original result and its several variables generalization (proved by Volkov [179] in the case when Ω is a compact set).

Corollary 5.1.2 *Suppose that Ω is a locally compact subset of the Euclidean space $\mathbb{R}^N$ and E is a vector sublattice of $\mathcal{F}(\Omega)$ that contains the test functions*

$$1,\ \mathrm{pr}_1, \ldots,\ \mathrm{pr}_N \ \textit{and} \sum_{k=1}^{N} \mathrm{pr}_k^2 .$$

If $(T_n)_n$ is a sequence of linear and positive operators from E into $\mathcal{F}(\Omega)$ such that

$$\lim_{n\to\infty} T_n(f) = f \quad \textit{uniformly on the compact subsets of } \Omega$$

for each of the $N+2$ aforementioned test functions, then the same property of convergence holds for all functions f in $E \cap C_b(\Omega)$.

For the convenience of the reader, we next mention a few examples of linear and positive operators obeying the hypotheses of Korovkin's Theorem or of its several variable extension provided by Corollary 5.1.2.

Example 5.1.3 The Bernstein–Kantorovich operators, $K_n : C([0,1]) \to C([0,1])$, were introduced by Kantorovich [104] via the formula

$$K_n(f)(x) = \sum_{k=0}^{n} \left[(n+1) \int_{k/n+1}^{(k+1)/n+1} f(t)\,\mathrm{d}t \right] \binom{n}{k} x^k (1-x)^{n-k}, \ n \geq 1,$$

and verifies the formulas

$$K_n(e_0)(x) = 1, \ K_n(e_1)(x) = \frac{nx}{n+1} + \frac{1}{2(n+1)}$$

and

$$K_n(e_2)(x) = \frac{n(n-1)x^2}{(n+1)^2} + \frac{2nx}{(n+1)^2} + \frac{1}{3(n+1)^2}.$$

According to Korovkin's Theorem, $K_n(f) \to f$ uniformly, for every f in $C([0,1])$. Moreover, due to the density of $C([0,1])$ in each of the Lebesgue spaces $L^p([0,1])$, with $1 \leq p < \infty$, one can easily show that also $K_n(f) \to f$ in $L^p([0,1])$. See Altomare [7, Theorem 3.10].

Example 5.1.4 The Szász–Mirakjan operators are defined by the formula

$$S_n(f)(x) = e^{-nx} \sum_{k=0}^{\infty} f\left(\frac{k}{n}\right) \frac{n^k x^k}{k!}$$

on various vector lattices of continuous functions for which the right-hand side series is absolutely convergent. If $f \in C_b([0,\infty))$, then $S_n(f) \to f$ uniformly on every compact subsets of $[0,\infty)$. Indeed,

$$S_n(e_0)(x) = 1, \ S_n(e_1)(x) = x \text{ and } S_n(e_2)(x) = x^2 + \frac{x}{n}.$$

Example 5.1.5 The Fejér convolution operators are defined by the formula

$$F_n(f)(x) = \frac{1}{2\pi} \int_{-\pi}^{\pi} \varphi_n(x-t) f(t)\mathrm{d}t$$

where

$$\varphi_n(x) = \begin{cases} \frac{\sin^2((n+1)x/2}{(n+1)\sin^2(x/2)}, & \text{if } x \text{ is not a multiple of } 2\pi \\ n+1, & \text{if } x \text{ is a multiple of } 2\pi. \end{cases}$$

In their case, $F_n(1) = 1$, $F_n(\cos) = \frac{n}{n+1}\cos$ and $F_n(\sin) = \frac{n}{n+1}\sin$, which leads to the fact that $F_n(f) \to f$ uniformly for $f \in C_{2\pi}(\mathbb{R})$ (and also in the p-mean for f in $L^p_{2\pi}(\mathbb{R})$, $1 \le p < \infty$).

Example 5.1.6 The multivariate Bernstein operators on the hypercube $[0,1]^N$,

$$B_n : C\left([0,1]^N\right) \to C\left([0,1]^N\right),$$

are defined by the formulas

$$\begin{aligned} &B_n(f)(x_1,\ldots,x_N) \\ &= \sum_{k_1=0}^{n} \cdots \sum_{k_N=0}^{n} p_{n,k_1}(x_1)\cdots p_{n,k_N}(x_N) f\left(\frac{k_1}{n_1},\ldots,\frac{k_N}{n_N}\right), \end{aligned}$$

where $p_{n,k_j}(x_j) = \binom{n}{k_j} x_j^{k_j}(1-x_j)^{n-k_j}$, $k_j \in \{1,\ldots,n\}$, $n \in \mathbb{N}$, and $x_j \in [0,1]$ for $j = 1,\ldots,N$. Since

$$B_n(1) = 1,\ B_n\left(\mathrm{pr}_k\right) = \mathrm{pr}_k \text{ and } B_n\left(\mathrm{pr}_k^2\right) = \frac{n-1}{n}\mathrm{pr}_k^2 + \frac{1}{n}\mathrm{pr}_k,$$

for every $k \in \{1,\ldots,N\}$, Corollary 5.1.2 applies and yields the Weierstrass Approximation Theorem for functions in $C\left([0,1]^N\right)$.

Example 5.1.7 Last, but not least, let us mention the case of the Gauss–Weierstrass convolution operators. They act on $C_b\left(\mathbb{R}^N\right)$ by the formulas

$$W_n(f)(\mathbf{x}) = \left(\frac{n}{4\pi}\right)^{N/2} \int_{\mathbb{R}^N} f(\mathbf{y}) e^{-n\|\mathbf{x}-\mathbf{y}\|^2/4}\, \mathbf{dy}.$$

See Weierstrass [184]. These operators verify the relations

$$W_n(1) = 1,\ W_n(\mathrm{pr}_k) = \mathrm{pr}_k \text{ and } W_n(\mathrm{pr}_k^2) = \mathrm{pr}_k^2 + 2/n$$

for every $k \in \{1,\ldots,N\}$. According to Corollary 5.1.2, $W_n(f) \to f$ uniformly on every compact subset of $\mathbb{R}^N$, whenever $f \in C_b\left(\mathbb{R}^N\right)$.

Let us pass now to the case of nonlinear operators. The one-dimensional case of Theorem 5.1.1 can be illustrated by the following three families of nonlinear operators:

– the *Bernstein–Kantorovich–Choquet operators*, which act on $C([0,1])$ by the formula

$$K_{n,\mu}(f)(x) = \sum_{k=0}^{n} \frac{(C)\int_{k/(n+1)}^{(k+1)/(n+1)} f(t)\,\mathrm{d}\mu}{\mu([k/(n+1),(k+1)/(n+1)])} \cdot \binom{n}{k} x^k (1-x)^{n-k};$$

– the *Szász–Mirakjan–Kantorovich–Choquet operators*, which act on appropriate subspaces of $C([0,\infty))$ by the formula

$$S_{n,\mu}(f)(x) = e^{-nx} \sum_{k=0}^{\infty} \frac{(C)\int_{k/n}^{(k+1)/n} f(t)\,\mathrm{d}\mu}{\mu([k/n,(k+1)/n])} \cdot \frac{(nx)^k}{k!};$$

– the *Baskakov–Kantorovich–Choquet operators*, which act on $C([0,\infty))$ by the formula

$$V_{n,\mu}(f)(x) = \sum_{k=0}^{\infty} \frac{(C)\int_{k/n}^{(k+1)/n} f(t)\,\mathrm{d}\mu}{\mu([k/n,(k+1)/n])} \cdot \binom{n+k-1}{k} \frac{x^k}{(1+x)^{n+k}}.$$

In all these cases, μ is the distorted measure

$$\mu = m^{1/2},$$

associated with the Lebesgue measure m (on suitable intervals of integration). Since μ is strongly subadditive (see Example 2.1.7), it follows that all the above operators are monotone increasing, sublinear, and comonotone additive.

The analysis of the behavior of these operators on the family of Korovkin test functions needs some computations.

Clearly, $K_{n,\mu}(e_0) = e_0$ and $K_{n,\mu}(1-e_1) = 1-e_1$. By Gal [68, Corollary 3.6 (i)], one can easily infer that $K_{n,\mu}(e_2) \to e_2$, uniformly on $[0,1]$. Hence, the operators $K_{n,\mu}$ $(n \in \mathbb{N})$ satisfy the hypotheses of Theorem 5.1.1.

In a similar manner, one can show that the operators $S_{n,\mu}$ and $V_{n,\mu}$ satisfy the hypotheses of Theorem 5.1.1 for $N = 1$ and $\Omega = [0,+\infty)$. In the first case, notice that the condition $S_{n,\mu}(e_0) = e_0$ is trivial. The convergence of the sequence of functions $S_{n,\mu}(e_2)(x)$ will be settled by computing the integrals

$$\begin{aligned} I_n &= \sqrt{n}\cdot(C)\int_{k/n}^{(k+1)/n} t^2\,\mathrm{d}\mu(t) \\ &= \sqrt{n}\int_0^{\infty} \mu(\{t \in [k/n,(k+1)/n] : t \geq \sqrt{\alpha}\})\,\mathrm{d}\alpha \end{aligned}$$

$$= \sqrt{n} \int_0^{((k+1)/n)^2} \mu(\{t \in [k/n, (k+1)/n] : t \geq \sqrt{\alpha}\}) \, \mathrm{d}\alpha$$

$$= \sqrt{n} \int_0^{(k/n)^2} \mu(\{t \in [k/n, (k+1)/n] : t \geq \sqrt{\alpha}\}) \, \mathrm{d}\alpha$$

$$+ \sqrt{n} \int_{(k/n)^2}^{((k+1)/n)^2} \mu(\{t \in [k/n, (k+1)/n] : t \geq \sqrt{\alpha}\}) \, \mathrm{d}\alpha.$$

We have

$$I_n = \sqrt{n} \cdot \left(\frac{k}{n}\right)^2 \cdot \frac{1}{\sqrt{n}} + \sqrt{n} \cdot \int_{(k/n)^2}^{((k+1)/n)^2} \sqrt{(k+1)/n - \sqrt{\alpha}} \, \mathrm{d}\alpha$$

$$= \left(\frac{k}{n}\right)^2 + 2\sqrt{n} \cdot \int_0^{1/n} \beta^{1/2}((k+1)/n - \beta) \, \mathrm{d}\beta$$

$$= \left(\frac{k}{n}\right)^2 + \sqrt{n} \cdot \frac{k+1}{n} \cdot \frac{4}{3} \cdot \beta^{3/2}\Big|_0^{1/n} - \sqrt{n} \cdot \frac{4}{5} \beta^{5/2}\Big|_0^{1/n}$$

$$= \left(\frac{k}{n}\right)^2 + \frac{4k+4}{3n^2} - \frac{4}{5n^2} = \frac{1}{15n^2}\left(15k^2 + 20k + 8\right),$$

whence

$$S_{n,\mu}(e_2)(x) = S_n(e_2)(x) + \frac{4}{3n} S_n(e_1)(x) + \frac{8}{15n^2} \to e_2(x)$$

as $n \to \infty$, uniformly in any compact subinterval $[0, a]$. Here $S_n = S_{n,m}$ is the classical Szász–Mirakjan–Kantorovich operator.

It remains to show that $S_{n,\mu}(-e_1)(x) \to -e_1(x)$, uniformly on any compact subinterval $[0, a]$. For this goal, we have to compute the following integral:

$$J_n = \sqrt{n} \cdot (C) \int_{k/n}^{(k+1)/n} (-t) \, \mathrm{d}\mu$$

$$= \sqrt{n} \int_{-\infty}^0 \left\{\mu(\{\omega \in [k/n, (k+1)/n] : -\omega \geq \alpha\}) - n^{-1/2}\right\} \mathrm{d}\alpha.$$

According to the definition of Choquet integral,

$$J_n = \sqrt{n} \int_{-k/n}^0 \left\{\mu(\{\omega \in [k/n, (k+1)/n] : \omega \leq -\alpha\}) - n^{-1/2}\right\} \mathrm{d}\alpha$$

$$+\sqrt{n}\int_{-(k+1)/n}^{-k/n}\left\{\mu(\{\omega\in[k/n,(k+1)/n]:\omega\le-\alpha\})-n^{-1/2}\right\}\mathrm{d}\alpha$$

$$=-\frac{k}{n}+\sqrt{n}\cdot\int_{-(k+1)/n}^{-k/n}\left(\sqrt{-\alpha-k/n}-n^{-1/2}\right)\mathrm{d}\alpha$$

$$=-\frac{k}{n}+\sqrt{n}\int_{k/n}^{(k+1)/n}\sqrt{\beta-k/n}\,\mathrm{d}\beta-\frac{1}{n}$$

$$=-\frac{k}{n}+\sqrt{n}\int_{0}^{1/n}\beta^{1/2}d\beta-\frac{1}{n}=-\frac{3k+1}{3n},$$

which implies that

$$S_{n,\mu}(-e_1)(x)=S_n(-e_1)(x)-\frac{1}{3n}\to-x\quad\text{as } n\to\infty,$$

uniformly on any compact interval $[0,a]$.

In a similar way, one can prove that the Baskakov–Kantorovich–Choquet operators $V_{n,\mu}$ $(n\in\mathbb{N})$ satisfy the hypotheses of Theorem 5.1.1.

An example involving several variables is that of bivariate Bernstein–Kantorovich–Choquet operators

$$K_{n,\mu}^{(2)}:C([0,1]^2)\to C([0,1]^2),$$

defined by the formula

$$K_{n,\mu}^{(2)}(f)(x_1,x_2)=\sum_{k_1=0}^{n}\sum_{k_2=0}^{n}p_{n,k_1}(x_1)p_{n,k_2}(x_2)$$

$$\cdot\frac{(\mathrm{C})\int_{k_1/(n+1)}^{(k_1+1)/(n+1)}\left((\mathrm{C})\int_{k_2/(n+1)}^{(k_2+1)/(n+1)}f(t_1,t_2)\mathrm{d}\mu(t_2)\right)\mathrm{d}\mu(t_1)}{\mu([k_1/(n+1),(k_1+1)/(n+1)])\mu([k_2/(n+1),(k_2+1)/(n+1)])},$$

where, as usually,

$$p_{n,k}(t)=\binom{n}{k}t^k(1-t)^{n-k}\quad\text{for } t\in[0,1] \text{ and } n\in\mathbb{N}$$

and

$$\mu=m^{1/2}$$

is the distorted measure associated to the Lebesgue measure on the specific intervals of integration.

Due to the properties of Choquet integral, it follows that each operator $K^{(2)}_{n,\mu}$ is weakly nonlinear, monotone increasing, and unital.

We will show that

$$K^{(2)}_{n,\mu}(f)(\mathbf{x}) \to f(\mathbf{x}) \quad \text{uniformly on } [0, 1]^2, \tag{5.3}$$

for all test functions $1, \ \pm \text{pr}_1, \ \pm \text{pr}_2, \ \text{pr}_1^2 + \text{pr}_2^2$ (which will imply, via Theorem 5.1.1, that this convergence occurs for all functions $f \in C([0, 1]^2)$).

The case of the unity is clear, while the case of the functions $\pm \text{pr}_1$ and $\pm \text{pr}_2$ is settled by the aforementioned properties of the operators $K^{(1)}_{n,\mu} = K_{n,\mu}$. As concerns the case of the function $\text{pr}_1^2 + \text{pr}_2^2$, notice that

$$
\begin{aligned}
&K^{(2)}_{n,\mu}(|\,\text{pr}_1^2 + \text{pr}_2^2 - \text{pr}_1^2(\mathbf{x}) - \text{pr}_2^2(\mathbf{x})|)(\mathbf{x})\\
&\leq \sum_{k=1}^{2} K^{(2)}_{n,\mu}(|\,\text{pr}_k + \text{pr}_k(\mathbf{x})| \cdot |\,\text{pr}_k - \text{pr}_k(\mathbf{x})|)(\mathbf{x})\\
&\leq 2\sum_{k=1}^{2} K^{(2)}_{n,\mu}(|\,\text{pr}_k - \text{pr}_k(\mathbf{x})|)(\mathbf{x}) \leq 2\sum_{k=1}^{2} \sqrt{K^{(2)}_{n,\mu}(|\,\text{pr}_k - \text{pr}_k(\mathbf{x})|^2)(\mathbf{x})}\\
&\leq 2\sum_{k=1}^{2} \sqrt{K^{(2)}_{n,\mu}(\text{pr}_k^2)(\mathbf{x}) + 2x_k K^{(2)}_{n,\mu}(-\,\text{pr}_k)(\mathbf{x}) + x_k^2},
\end{aligned}
$$

according to Hölder's type inequality stated by Theorem 4.6.1. Now, by using the calculations for the Bernstein–Kantorovich–Choquet operators in one variable, it is immediate that $K^{(2)}_{n,\mu}(\text{pr}_k^2)(\mathbf{x}) \to x_k^2$ and $K^{(2)}_{n,\mu}(-\,\text{pr}_k)(\mathbf{x}) \to -x_k$ as $n \to \infty$, uniformly with respect to $\mathbf{x} = (x_1, x_2) \in [0, 1]^2$, for $k \in \{1, 2\}$. Therefore,

$$K^{(2)}_{n,\mu}(\text{pr}_k^2)(\mathbf{x}) + 2x_k K^{(2)}_{n,\mu}(-\,\text{pr}_k)(\mathbf{x}) + x_k^2 \to 0,$$

uniformly with respect to $\mathbf{x}$, for $k \in \{1, 2\}$. Thus, the convergence 5.3 also occurs for the function $\text{pr}_1^2 + \text{pr}_2^2$ (and therefore for all functions $f \in C([0, 1]^2)$.

Remark 5.1.8 (The Necessity of Hypotheses in Theorem 5.1.1) Though being sufficient for the fulfillment of Theorem 5.1.1, none of the three conditions imposed on the operators T_n (sublinearity, monotonicity, and translatability) is necessary. See the case of the sequence of operators,

$$T_n : C([0, 1]) \to C([0, 1]), \quad T_n(f) = f + f^2/n,$$

which fails all these assumptions despite its uniform convergence to the identity of $C([0, 1])$.

5.2 Separating Functions

The alert reader has probably already noticed that the basic clue in the proof of Theorem 5.1.1 is the estimate (5.2), characterized in [134] as a property of absolute continuity. See also [132] and [133]. This estimate occurs in the larger context of spaces $C(K)$, where K is a compact metric space on which there is defined a *separating function*, i.e., a nonnegative continuous function $\gamma : K \times K \to \mathbb{R}$ such that

$$\gamma(x, y) = 0 \text{ implies } x = y.$$

For example, if $f_1, \ldots, f_m \in C(K)$ is a family of functions that separates the points of K, then

$$\gamma(x, y) = \sum_{k=1}^{m} (f_k(x) - f_k(y))^2 \tag{5.4}$$

is a separating function. The separating function associated to the family of coordinate functions $\mathrm{pr}_1, \ldots, \mathrm{pr}_N$ in $\mathbb{R}^N$ is the square distance,

$$\gamma(\mathbf{x}, \mathbf{y}) = \|\mathbf{x} - \mathbf{y}\|^2 .$$

If ψ is a separating function and $\varphi : [0, \infty) \to [0, \infty)$ is a function such that $\varphi(\alpha) = 0$ only for $\alpha = 0$, then $\gamma(x, y) = \varphi(\psi(x, y))$ is also a separating function.

Lemma 5.2.1 (Lomelí and García [119] and Niculescu [134]) *If K is a compact metric space, and $\gamma : K \times K \to \mathbb{R}$ is a separating function, then for every real-valued continuous function f defined on K and every $\varepsilon > 0$, there exists $\delta(\varepsilon) > 0$ such that*

$$|f(x) - f(y)| \leq \varepsilon + \delta(\varepsilon)\gamma(x, y) \quad \textit{for all } x, y \in K.$$

Proof If the estimate above doesn't work, then for a suitable $\varepsilon_0 > 0$, one can find two sequences $(x_n)_n$ and $(y_n)_n$ of elements of K such that

$$|f(x_n) - f(y_n)| \geq \varepsilon_0 + 2^n \gamma(x_n, y_n) \tag{5.5}$$

for all n. Without loss of generality, we may assume (by passing to subsequences) that both sequences $(x_n)_n$ and $(y_n)_n$ are convergent, respectively to x and y. Since f is bounded, the inequality (5.5) forces $x = y$. Indeed,

$$\frac{|f(x_n) - f(y_n)|}{2^n} \to 0 \text{ and } \frac{|f(x_n) - f(y_n)|}{2^n} \geq \gamma(x_n, y_n) \to \gamma(x, y) \geq 0,$$

which implies that $\gamma(x, y) = 0$. On the other hand, from (5.5), one can infer that $|f(x) - f(y)| \geq \varepsilon_0$ and thus, $x \neq y$. This contradiction shows that the assumption made at the beginning of the proof is wrong and the assertion of Lemma 5.2.1 is true. ∎

The argument of Lemma 5.2.1 also shows that every separating function $\gamma : K \times K \to \mathbb{R}$ is related to the metric d on K via an estimate of the form

$$d(x, y) \leq \varepsilon + \delta(\varepsilon)\gamma(x, y) \quad \text{for all } (x, y) \in K \times K \text{ and } \varepsilon > 0.$$

Every separating function $\gamma : K \times K \to \mathbb{R}$ generates a family of nonnegative continuous functions on K, precisely,

$$\gamma_x : K \to \mathbb{R}, \quad \gamma_x(y) = \gamma(x, y) \text{ for } x, y \in K.$$

As shows Theorem 5.2.2 below, this family can be used as a family of test functions in the same manner as the functions 1, x and x^2 were used in Korovkin's theorem.

The following result represents a nonlinear generalization of Korovkin's theorem as well as of many other related results existing in the literature.

Theorem 5.2.2 *Let K be a compact metric space (endowed with the metric d), and let $(T_n)_n$ be a sequence of sublinear and monotone increasing operators from $C(K)$ into itself such that*

$$T_n(1) \to 1 \quad \textit{uniformly on } K. \tag{5.6}$$

Suppose that $\gamma : K \times K \to \mathbb{R}$ is a separating function such that

$$T_n(\gamma_x)(x) \to 0 \quad \textit{uniformly on } K. \tag{5.7}$$

Then for all nonnegative functions $f \in C(K)$,

$$T_n(f) \to f \quad \textit{uniformly on } K. \tag{5.8}$$

This convergence occurs for all functions $f \in C(K)$ if the operators T_n are weakly nonlinear and monotone increasing.

Proof Let $f \in C(K)$ be a nonnegative function. Then, according to Lemma 5.2.1, for every $\varepsilon > 0$, there is $\delta(\varepsilon) > 0$ such that

$$|f - f(x)| \leq \varepsilon + \delta(\varepsilon)\gamma_x \tag{5.9}$$

for all $x \in K$. Since the operators T_n are sublinear and monotone increasing, it follows that

$$\begin{aligned} |T_n(f) - f|(x) &\leq |T_n(f)(x) - f(x)T_n(1)(x)| + |f(x)(T_n(1)(x) - 1)| \\ &= |T_n(f) - T_n(f(x))|(x) + |f(x)(T_n(1)(x) - 1)| \\ &\leq T_n(|f - f(x)|)(x) + |f(x)(T_n(1)(x) - 1)| \\ &\leq \varepsilon T_n(1)(x) + \delta(\varepsilon)T_n(\gamma_x)(x) + |f(x)(T_n(1)(x) - 1)|. \end{aligned}$$

According to our hypotheses (5.6) and (5.7), this leads to the conclusion that $T_n(f) \to f$, uniformly on K.

Suppose now that $f \in C(K)$ is an arbitrary function and each operator T_n is also translatable. Then $T_n(f+\|f\|_\infty) = T_n(f)+\|f\|_\infty \cdot T_n(1)$ while, according to the previous considerations,

$$T_n(f + \|f\|_\infty) \to f + \|f\|_\infty.$$

Taking into account the hypothesis (5.6), we conclude that $T_n(f) \to f$ in $C(K)$. ■

An estimate of the convergence stated in Theorem 5.2.2 is as follows.

Theorem 5.2.3 *Suppose that K is a compact metric space endowed with the metric d and $(T_n)_n$ is a sequence of weakly nonlinear and monotone increasing operators from $C(K)$ into itself such that $T_n(1) = 1$ for every $n \in \mathbb{N}$. Then every Lipschitz function $f \in C(K)$ verifies the estimate*

$$|T_n(f)(x) - f(x)| \leq \|f\|_{\text{Lip}} \cdot \sup\left\{\left|T_n(d_y^2)(y)\right|^{1/2} : y \in K\right\}$$

whenever $x \in K$ and $n \in \mathbb{N}$.

Here d_y denotes the function defined on K by the formula $d_y(x) = d(x, y)$.

Proof According to the Lipschitz property,

$$f(x) \cdot 1 - \alpha d_x \leq f \leq f(x) \cdot 1 + \alpha d_x \quad \text{for every } x \in K, \tag{5.10}$$

where $\alpha = \|f\|_{\text{Lip}}$. Suppose for the moment that $f \geq 0$. Since the operators T_n are sublinear and monotone increasing, one can infer from the left-hand side inequality (rewritten as $f(x) \cdot 1 \leq f + \alpha d_x$), the relation

$$f(x) \le T_n(f)(x) + \alpha T_n(d_x)(x).$$

Applying the operators T_n to the right-hand side of (5.10), one obtains

$$T_n(f)(x) \le f(x) + \alpha T_n(d_x)(x).$$

Then

$$|T_n(f)(x) - f(x)| \le \alpha T_n(d_x)(x) \le \alpha \left(T_n(d_x^2)(x)\right)^{1/2}$$

for every $x \in K$ and $n \in \mathbb{N}$, where the second inequality is motivated by the generalization of Hölder's inequality provided by Theorem 4.6.1.

The case of real-valued Lipschitz functions can be settled via the identity

$$|T_n(f)(x) - f(x)| = |T_n(f + \|f\| \cdot 1)(x) - (f(x) + \|f\| \cdot 1)|$$

and the property of translatability of the operators T_n. ∎

Remark 5.2.4 In the case of a separating function of the form

$$\gamma(x, y) = \sum_{k=1}^{p} |f_k(x) - f_k(y)|^2,$$

associated with a finite family $f_1, \ldots, f_p$ of continuous functions that separates the points of the compact metric space K, one can prove the following variant of Theorem 5.2.2:

If $(T_n)_n$ is a sequence of sublinear and monotone increasing operators from $C(K)$ into itself such that

$$T_n(f) \to f \quad \text{uniformly on } K,$$

for each of the test functions $1, \ \pm f_1, \ldots, \ \pm f_p$ and $\sum_{k=1}^{2} f_k^2$, then for all nonnegative functions $f \in C(K)$,

$$T_n(f) \to f \quad \text{uniformly on } K.$$

This convergence occurs for all functions $f \in C(K)$ if the operators T_n are weakly nonlinear and monotone increasing.

The proof can be done by adapting the argument of Theorem 5.2.2.

Notice that Remark 5.2.4 represents a partial generalization of Theorem 5.1.1. Several examples illustrating Remark 5.2.4 are presented in what follows.

Consider first the case of the unit circle

$$S^1 = \{(\cos\varphi, \sin\varphi) : \varphi \in \mathbb{R}\},$$

endowed with the metric induced by $\mathbb{R}^2$. S^1 is a compact (, i.e., bounded and close) set of $\mathbb{R}^2$ and the functions $\mathrm{pr}_1, \mathrm{pr}_2 : S^1 \to \mathbb{R}$ constitute a separating family for it. As is well known, the Banach lattice $C(S^1)$ can be identified with the Banach lattice $C_{2\pi}(\mathbb{R})$, of all continuous and 2π-periodic functions $f : \mathbb{R} \to \mathbb{R}$. By this identification, the family $\{\mathrm{pr}_1, \mathrm{pr}_2\}$ corresponds to the family $\{\cos, \sin\}$ and, according to Remark 5.2.4, a sufficient condition for a sequence of weakly nonlinear and monotone increasing operators $T_n : C_{2\pi}(\mathbb{R}) \to C_{2\pi}(\mathbb{R})$ to verify the condition

$$\lim_{n\to\infty} T_n(f) \to f \quad \text{uniformly on } \mathbb{R},$$

is to verify this condition for the test functions $1, \ \pm\cos\ , \ \pm\sin$. In the case of sequences of linear and positive operators, the family of test functions can be reduced to $1, \ \cos$ and $\sin$ (a result noticed by Korovkin [107, 108]).

Consider now the case of product spaces $K \times S^1$, where K is a compact metric space. The space $C(K \times S^1)$ can be identified with the space $C_{2\pi}(K \times \mathbb{R})$, of all continuous functions $f : K \times \mathbb{R} \to \mathbb{R}$, 2π-periodic in the second variable, endowed with the sup norm. This space is genuine for many results in dynamical systems theory.

Popa [152] has proved the variant of Remark 5.2.4 for the linear and positive operators $T_n : C(K \times S^1) \to C(K \times S^1)$ with $K \subset \mathbb{R}^N$. The reader can easily check that his result also holds in the setting of weakly nonlinear and monotone increasing operators. Precisely, the convergence

$$T_n(f) \to f \quad \text{uniformly on } K \times S^1,$$

for every $f \in C(K \times S^1)$ reduces to its validity for the product functions $f(\mathbf{x}, \varphi) = u(\mathbf{x})v(\varphi)$, where

$$u \in \left\{1, \ \pm\mathrm{pr}_1, \ldots, \ \pm\mathrm{pr}_N, \sum_{k=1}^{N} \mathrm{pr}_k^2\right\} \text{ and } v \in \{1, \ \pm\cos\ , \ \pm\sin\}.$$

Some other cases of interest (like the torus $S^1 \times S^1$ and the 2-dimensional sphere S^2) are left to the reader as an exercise.

Remark 5.2.5 In the absence of the condition of translatability, the conclusion of Theorem 5.2.2 (as well as of Remark 5.2.4) may fail for functions with variable sign. An example working for $K = [0, 1]$ is given by the Bernstein like operators

$$T_n(f)(x) = \sum_{k=0}^{n} \binom{n}{k} x^k (1-x)^{n-k} f^+\left(\frac{k}{n}\right),$$

which are sublinear and monotone increasing (but not translatable). By Theorem 5.2.2, applied for the separating function $\gamma(x, y) = (x-y)^2$, $T_n(f) \to f$ uniformly on $[0, 1]$ for all nonnegative functions f in $C([0, 1])$. However, this fails for the real-valued functions. Remarkably, the case of nonnegative functions is strong enough to provide valuable information for *all* functions in $C([0, 1])$, for example, the possibility to approximate them by polynomials. Indeed,

$$T_n(f + \|f\|_\infty) - \|f\|_\infty \to f \quad \text{uniformly on } [0, 1].$$

5.3 The Extension of a Result Due to Altomare

The usefulness of the technique of separating functions is also supported by the following nonlinear analog of a result due to Altomare (see [7, Theorem 3.5, p. 100]).

Theorem 5.3.1 *Let M be a locally compact metric space (endowed with the metric d) and consider a vector sublattice E of $\mathcal{F}(M)$ containing the constant functions and all the functions $d_x^p = d^p(\cdot, x)$, for $x \in M$ and some exponent $p \in [1, \infty)$. Let $(T_n)_n$ be a sequence of sublinear and monotone increasing operators from E into $\mathcal{F}(M)$, which verifies the following two conditions:*

(i) $\lim_{n\to\infty} T_n(1) = 1$, *uniformly on the compact subsets of M;*
(ii) $\lim_{n\to\infty} T_n(d_x^p)(x) = 0$, *uniformly on the compact subsets of M;*
Then, for all nonnegative f in $E \cap C_b(M)$, we have

$$\lim_{n\to\infty} T_n(f) = f \quad \textit{uniformly on the compact subsets of } M.$$

The convergence occurs for all functions in $E \cap C_b(M)$ when the operators T_n are also translatable.

The proof of Theorem 5.3.1, needs the following topological lemma, also due to Altomare. See [7, Lemma 3.4, p. 99], for details.

Lemma 5.3.2 *Let M be as in the statement of Theorem 5.3.1. Then for every compact subset K of M and for every $\varepsilon > 0$, there exist $0 < \varepsilon' < \varepsilon$ and a compact subset K_ε of M such that the open ball $B_{\varepsilon'}(x)$ is included in K_ε for every $x \in K$.*

Proof of Theorem 5.3.1 Let $f \in E \cap C_b(M)$ and $\varepsilon > 0$ be arbitrarily fixed. Then for every compact subset K of M, choose $\varepsilon' \in (0, \varepsilon)$ and K_ε as in Lemma 5.3.2. Since f is uniformly continuous on K_ε, there exists $\delta \in (0, \varepsilon')$ such that

$$|f(x) - f(y)| \leq \varepsilon \quad \text{for every } x, y \in K_\varepsilon \text{ with } d(x, y) \leq \delta.$$

Suppose that $x \in K$ and $y \in M$. If $d(x, y) \leq \delta$, then $y \in B'(x, \overline{\varepsilon}) \subset K_\varepsilon$ and therefore, $|f(x) - f(y)| \leq \varepsilon$. If $d(x, y) \geq \delta$, then

$$|f(x) - f(y)| \leq \frac{2\|f\|_\infty}{\delta^p} \cdot d^p(x, y).$$

Therefore,

$$|f - f(x)| \leq \frac{2\|f\|_\infty}{\delta^p} \cdot d_x^p + \varepsilon \cdot 1 \quad \text{for every } x \in K,$$

so that by applying T_n to this inequality we infer in the case of nonnegative functions f that

$$\begin{aligned} |T_n(f)(x) - f(x)T_n(1)(x)| &\leq T_n(|f - f(x)|)(x) \\ &\leq \frac{2\|f\|_\infty}{\delta^p} \cdot T_n(d_x^p)(x) + \varepsilon T_n(1)(x) \end{aligned} \tag{5.11}$$

for all $x \in K$. See Remark 4.1.14. As a consequence,

$$\lim_{n\to\infty} T_n(f)(x) = f(x) \quad \text{uniformly on } K. \tag{5.12}$$

Assume now that all operators T_n are also translatable and f is an arbitrary function in $E \cap C_b(M)$. Then the left-hand side inequality in (5.11) still works (because it works for $f + \|f\|_\infty$,), so the inequality (5.11) and its consequence (5.12) remain valid for every $f \in E \cap C_b(M)$. The proof is done. ■

Several illustrations of Theorem 5.3.1 for $M \subset \mathbb{R}^2$, endowed with the metric

$$d(\mathbf{x}, \mathbf{y}) = |x_1 - y_1| + |x_2 - y_2|, \tag{5.13}$$

(equivalent to the Euclidean metric), are presented in what follows.

Example 5.3.3 (The Bivariate Possibilistic Bernstein–Durrmeyer and Kantorovich Operators) These operators act on $C([0, 1]^2)$ respectively by the formulas

$$P_n(f)(x_1, x_2) = \sum_{k_1=0}^{n}\sum_{k_2=0}^{n} p_{n,k_1}(x_1)p_{n,k_2}(x_2)$$
$$\cdot \frac{\sup\{f(t_1, t_2)t_1^{k_1}(1-t_1)^{n-k_1}t_2^{k_2}(1-t_2)^{n-k_2} : t_1, t_2 \in [0, 1]\}}{k_1^{k_1}n^{-n}(n-k_1)^{n-k_1}k_2^{k_2}n^{-n}(n-k_2)^{n-k_2}},$$

and

$$Q_n(f)(x_1, x_2) = \sum_{k_1=0}^{n}\sum_{k_2=0}^{n} p_{n,k_1}(x_1)p_{n,k_2}(x_2)$$
$$\cdot \sup\left\{f(t_1, t_2) : t_1 \in \left[\frac{k_1}{n+1}, \frac{k_1+1}{n+1}\right], t_2 \in \left[\frac{k_2}{n+1}, \frac{k_2+1}{n+1}\right]\right\},$$

where

$$p_{n,k}(x) = \binom{n}{k}x^k(1-x)^{n-k}.$$

They are obtained from the classical bivariate operators of Bernstein–Durrmeyer and Kantorovich by replacing the Lebesgue integral by the integral associated to a possibility measure. See the paper of Gal [70].

It is easy to show that both operators P_n and Q_n are monotone increasing, unital, and sublinear. Notice that Q_n is translatable, while P_n is not.

Using the estimate included in the proof of Corollary 3.5 in Gal [70] (for functions of one variable), one can easily show that

$$P_n(|\,\mathrm{pr}_k - \mathrm{pr}_k(\mathbf{x})|)(\mathbf{x})$$
$$\leq \frac{(1+\sqrt{2})\sqrt{\mathrm{pr}_k(\mathbf{x})(1-\mathrm{pr}_k(\mathbf{x}))} + \sqrt{2}\sqrt{\mathrm{pr}_k(\mathbf{x})}}{\sqrt{n}} + \frac{1}{n}, \tag{5.14}$$

for $k \in \{1, 2\}$, $n \in \mathbb{N}$, and all points $\mathbf{x} = (x_1, x_2)$ and $\mathbf{t} = (t_1, t_2)$ in $[0, 1]^2$. Taking into account Theorem 5.3.1 (for $p = 1$ and the metric 5.13), we conclude that

$$P_n(f)(\mathbf{x}) \to f(\mathbf{x}), \quad \text{uniformly on } [0, 1]^2,$$

for all nonnegative functions $f \in C([0, 1]^2)$.

We arrive at the same conclusion by applying Theorem 5.1.1. According to this theorem, we have to show that $P_n(f) \to f$ uniformly on $[0, 1]^2$ for all test functions $1, \pm\mathrm{pr}_1, \pm\mathrm{pr}_2, \mathrm{pr}_1^2 + \mathrm{pr}_2^2$. As in the case of Bernstein–Kantorovich–Choquet operators, only the status of the test function $\mathrm{pr}_1^2 + \mathrm{pr}_2^2$ needs attention. Indeed,

$$
\begin{aligned}
&P_n(|\operatorname{pr}_1^2+\operatorname{pr}_2^2-\operatorname{pr}_1^2(\mathbf{x})-\operatorname{pr}_2^2(\mathbf{x})|)\\
&\leq\sum_{k=1}^{2}P_n(|\operatorname{pr}_k)+\operatorname{pr}_k(\mathbf{x})|\cdot|\operatorname{pr}_k-\operatorname{pr}_k(\mathbf{x})|)\\
&\leq 2\sum_{k=1}^{2}P_n(|\operatorname{pr}_k)-\operatorname{pr}_k(\mathbf{x})|)(x),
\end{aligned}
$$

so that, according to (5.14), we infer that $P_n(\operatorname{pr}_1^2+\operatorname{pr}_2^2)\rightarrow\operatorname{pr}_1^2+\operatorname{pr}_2^2$, uniformly on $[0,1]^2$.

The case of operators Q_n is similar. The fulfilment of the hypotheses of 5.3.1 (for $p=1$ and the metric 5.13), can be shown as above, by using the following estimate included in the proof of Theorem 3.7 in Gal [70]:

$$
Q_n(|\operatorname{pr}_k-\operatorname{pr}_k(\mathbf{x})|)(\mathbf{x})\leq\frac{\sqrt{\operatorname{pr}_k(\mathbf{x})(1-\operatorname{pr}_k(\mathbf{x}))}}{\sqrt{n}}+\frac{2}{n+1}.
$$

Since the operators Q_n are translatable, the convergence of $Q_n(f)$ to f holds for all $f\in C([0,1]^2)$.

Example 5.3.4 (The Case of Max-Product Operators) An important class of monotone increasing, unital, and sublinear operators are the so-called max-product operators, whose theory can be found in the monograph of Bede, Coroianu and Gal [20]. Denote

$$
\Delta^{(2)}=\{(x_1,x_2):x_1,x_2\geq 0,x_1+x_2\leq 1\},
$$

the 2-dimensional simplex. The *max-product operators* $T_n:C(\Delta^{(2)})\rightarrow C(\Delta^{(2)})$, of Bernstein type, are defined by the formula

$$
T_n(f)(x_1,x_2)=\frac{\bigvee_{i=0}^{n}\bigvee_{j=0}^{n-i}\binom{n}{i}\binom{n-i}{j}x_1^ix_2^j(1-x_1-x_2)^{n-i-j}f(i/n,j/n)}{\bigvee_{i=0}^{n}\bigvee_{j=0}^{n-i}\binom{n}{i}\binom{n-i}{j}x_1^ix_2^j(1-x_1-x_2)^{n-i-j}}.
$$

Here $\bigvee$ stands for maximum.

As shown by Bede, Coroianu, and Gal [20, pp. 139–140], these max-product operators satisfy the estimate

$$
T_n(|\operatorname{pr}_k-\operatorname{pr}_k(\mathbf{x})|)(\mathbf{x})\leq\frac{6}{\sqrt{n+1}}
$$

for $k\in\{1,2\},\ n\in\mathbb{N}$, and all points $\mathbf{x}=(x_1,x_2)\in\Delta^{(2)}$.

The operators T_n are sublinear, unital, and monotone increasing (but not translatable). Taking into account Theorem 5.3.1 (for $p = 1$ and the metric 5.13), we conclude that $T_n(f)(\mathbf{x}) \to f(\mathbf{x})$ uniformly on $\Delta^{(2)}$ for all nonnegative functions $f \in C(\Delta^{(2)})$.

5.4 Korovkin-Type Theorems for Almost Everywhere Convergence

A useful remark concerning Theorem 5.1.1 is its validity in various other contexts of convergence like convergence almost everywhere, convergence in measure, convergence in mean, etc. The case of almost everywhere convergence in the sense of Lebesgue measure is detailed in what follows.

$\mathcal{AC}_b(\Omega)$ denotes the vector lattice of all almost everywhere continuous and bounded functions $f : \Omega \to \mathbb{R}$.

Theorem 5.4.1 *Suppose that Ω is a locally compact subset of the Euclidean space $\mathbb{R}^N$ endowed with a positive Borel measure μ and E is a vector sublattice of $\mathcal{F}(\Omega)$ that contains the test functions $1, \ \pm \mathrm{pr}_1, \ldots, \ \pm \mathrm{pr}_N$ and $\sum_{k=1}^N \mathrm{pr}_k^2$.*

(i) *If $(T_n)_n$ is a sequence of sublinear and monotone increasing operators from E into E such that*

$$T_n(f)(\mathbf{x}) \to f(\mathbf{x}) \quad a.e.$$

for each of the $2N+2$ aforementioned test functions, then this property extends to all nonnegative functions f in $E \cap \mathcal{AC}_b(\Omega)$.

(ii) *If, in addition, each operator T_n is translatable, then $T_n(f)(\mathbf{x}) \to f(\mathbf{x})$ a.e. for every $f \in E \cap \mathcal{AC}_b(\Omega)$.*

(iii) *When Ω is included in the positive cone of $\mathbb{R}^N$, then, in both cases* (i) *and* (ii)*, the family of test functions can be reduced to $1, \ -\mathrm{pr}_1, \ldots, \ -\mathrm{pr}_N$ and $\sum_{k=1}^N \mathrm{pr}_k^2$.*

Proof

(i) Let $f \in E \cap \mathcal{AC}_b(\Omega)$, and let ω be a point of continuity of f, which is also a point where

$$T_n(h)(\omega) \to h(\omega)$$

for each of the functions $h \in \left\{1, \ \pm \mathrm{pr}_1, \ldots, \ \pm \mathrm{pr}_N \text{ and } \sum_{k=1}^N \mathrm{pr}_k^2\right\}$.

Then for $\varepsilon > 0$ arbitrarily fixed, there is $\delta > 0$ such that

$$|f(\mathbf{x}) - f(\omega)| \leq \varepsilon \quad \text{for every } \mathbf{x} \in \Omega \text{ with } \|\mathbf{x} - \omega\| \leq \delta.$$

If $\|\mathbf{x} - \omega\| \geq \delta$, then

$$|f(\mathbf{x}) - f(\omega)| \leq \frac{2\|f\|_\infty}{\delta^2} \cdot \|\mathbf{x} - \omega\|^2,$$

so that

$$|f(\mathbf{x}) - f(\omega)| \leq \varepsilon + \frac{2\|f\|_\infty}{\delta^2} \cdot \|\mathbf{x} - \omega\|^2 \quad \text{for every } \mathbf{x} \in \Omega. \tag{5.15}$$

Denoting

$$M = \max\left\{\mathrm{pr}_1(\omega), \ldots, \mathrm{pr}_N(\omega), 0\right\},$$

one can restate (5.15) as

$$|f(\mathbf{x}) - f(\omega)| \leq \varepsilon + \frac{2\|f\|_\infty}{\delta^2} \left[\sum_{k=1}^N \mathrm{pr}_k^2(\mathbf{x}) + 2\sum_{k=1}^N \mathrm{pr}_k(\mathbf{x})(M - \mathrm{pr}_k(\omega)) \right.$$
$$\left. +2M \sum_{k=1}^N \left(-\mathrm{pr}_k(\mathbf{x})\right) + \|\omega\|^2 \right].$$

Taking into account the properties of sublinearity and monotonicity of the operators T_n, we infer in the case where $f \geq 0$ that

$$\begin{aligned}
|T_n(f)(\omega) - f(\omega)| &\leq |T_n(f)(\omega) - T_n(f(\omega) \cdot 1)(\omega) + f(\omega)T_n(1)(\omega) - f(\omega)| \\
&\leq T_n(|f - f(\omega)|)(\omega) + f(\omega)|T_n(1)(\omega) - 1|
\end{aligned}$$
$$\leq \varepsilon T_n(1)(\omega) + \frac{2\|f\|_\infty}{\delta^2} \left[T_n\left(\sum_{k=1}^N \mathrm{pr}_k^2\right)(\omega) + 2\sum_{k=1}^N (M - \mathrm{pr}_k(\omega)) T_n\left(\mathrm{pr}_k\right)(\omega) \right.$$
$$\left. +2M \sum_{k=1}^N T_n\left(-\mathrm{pr}_k\right)(\omega) + \|\omega\|^2\, T_n(1)(\omega) \right] + f(\omega)|T_n(1)(\omega) - 1|.$$

By our choice of ω,

$$\limsup_{n\to\infty} |T_n(f)(\boldsymbol{\omega}) - f(\omega)| \leq \varepsilon,$$

whence we conclude that $T_n(f)(\omega) \to f(\boldsymbol{\omega})$, since $\varepsilon > 0$ was arbitrarily fixed.

(ii) Suppose in addition that each operator T_n is also translatable. According to the assertion (i),

$$T_n(f + \|f\|_\infty) \to f + \|f\|_\infty \quad \text{a.e.}$$

Since the operators T_n are translatable, $T_n(f + \|f\|_\infty) = T_n(f) + \|f\|_\infty T_n(1)$ and by our hypotheses $T_n(1) \to 1$ a.e. Therefore, $T_n(f) \to f$ a.e.
As concerns the assertion (iii), notice that when Ω is included in the positive cone of $\mathbb{R}^N$, one can restate the estimate (5.15) as

$$|f(\mathbf{x}) - f(\omega)| \leq \varepsilon + \frac{2\|f\|_\infty}{\delta^2}\left[\sum_{k=1}^{N} \mathrm{pr}_k^2(\mathbf{x}) + 2\sum_{k=1}^{N}(-\mathrm{pr}_k(\mathbf{x}))\,\mathrm{pr}_k(\omega)) + \|\omega\|^2\right],$$

which leads to

$$\begin{aligned}|T_n(f)(\omega) - f(\omega)| &\leq T_n(|f - f(\boldsymbol{\omega})|)(\omega) + f(\omega)|T_n(1)(\omega) - 1| \\ &\leq \varepsilon + \frac{2\|f\|_\infty}{\delta^2} \\ &\cdot \left[T_n\left(\sum_{k=1}^{N} \mathrm{pr}_k^2\right)(\omega) + 2\sum_{k=1}^{N} \mathrm{pr}_k(\omega) T_n\left(-\mathrm{pr}_k\right)(\omega) + \|\omega\|^2\, T_n(1)(\omega)\right] \\ &\quad + f(\omega)|T_n(1)(\omega) - 1|\end{aligned}$$

in the case where $f \geq 0$. Then the proof continues verbatim as in the cases (i) and (ii). ■

Corollary 5.4.2 *Let $\mathcal{R}(K)$ be the vector lattice of all Riemann integrable functions defined on an N-dimensional compact interval $K = \prod_{k=1}^{N}[a_k, b_k]$, and let $(T_n)_n$ be a sequence of sublinear and monotone operators from $\mathcal{R}(K)$ into itself such that*

$$T_n(f) \to f \quad a.e.$$

for each of the functions $1, \ \pm\mathrm{pr}_1, \ldots, \ \pm\mathrm{pr}_N$ and $\sum_{k=1}^{N} \mathrm{pr}_k^2$. Then this convergence also occurs for all nonnegative functions $f \in \mathcal{R}(K)$. It occurs for all Riemann integrable functions defined on K when the operators T_n are weakly nonlinear and monotone increasing.

Proof One applies Theorem 5.4.1, by taking into account Lebesgue's characterization of Riemann integrable functions: a function $f : K \to \mathbb{R}$ belongs to $\mathcal{R}(K)$ if and only if f is bounded and the set of points where it is not continuous has Lebesgue measure zero. See Choudary-Niculescu [43, p. 323], for the case $N = 1$ and Zorich [196, pp. 110–113, section 11.1], for $N \geq 1$. ■

Remark 5.4.3 Suppose that E is a vector sublattice of $\mathcal{F}(\Omega)$ as in Theorem 5.4.1, and let $(T_n)_n$ be a sequence of weakly nonlinear and monotone increasing operators from E into $\mathcal{F}(\Omega)$. An inspection of the argument of Theorem 5.4.1 reveals the following fact. If $f \in E \cap \mathcal{F}_b(\Omega)$ and Ω_f is a subset of its set of points of continuity, then

$$T_n(f)(\mathbf{x}) \to f(\mathbf{x}) \quad \text{for every } \mathbf{x} \in \Omega_f,$$

provided that

$$T_n(h)(\mathbf{x}) \to h(\mathbf{x}) \quad \text{for every } \mathbf{x} \in \Omega_f,$$

whenever $h \in \{1, \pm\,\mathrm{pr}_1, \ldots, \pm\,\mathrm{pr}_N, \sum_{k=1}^N \mathrm{pr}_k^2\}$.

This remark also applies in the case of Corollary 5.4.2.

Consider the particular case of *Bernstein–Kantorovich operators,*

$$K_n(f)(x) = (n+1)\sum_{k=0}^{n} p_{n,k}(x) \cdot \int_{k/(n+1)}^{(k+1)/(n+1)} f(t)\mathrm{d}t,$$

viewed as operators from $\mathcal{R}([0,1])$ into itself.

As was already noticed, $K_n(e_k) \to e_k$ uniformly on $[0,1]$ for $k \in \{0,1,2\}$, which implies that the same happens for the test function $-e_1$. According to Remark 5.4.3, we infer the following result previously noticed by Lorentz [120, Theorem 2.1.1, p. 30]:

Theorem 5.4.4 *$K_n(f)(x) \to f(x)$ at each point of continuity of $f \in \mathcal{R}([0,1])$ (and thus, $K_n(f)(x) \to f(x)$ a.e. on $[0,1]$).*

In connection with Remark 5.4.3, let us mention the following nonlinear generalization of a result due to Altomare. See [8, Remark 2.4].

Theorem 5.4.5 *Suppose that Ω is a subset of the Euclidean space $\mathbb{R}^N$, ω is a point in Ω and E is a vector sublattice of $\mathcal{F}(\Omega)$ that contains the test functions $1, \pm\,\mathrm{pr}_1, \ldots, \pm\,\mathrm{pr}_N$ and $\sum_{k=1}^N \mathrm{pr}_k^2$.*

If $(T_n)_n$ is a sequence of sublinear and monotone increasing functionals defined on E such that

$$T_n(f) \to f(\omega)$$

whenever f *is one of the test functions* $1, \pm \mathrm{pr}_1, \ldots, \pm \mathrm{pr}_N$ *and* $\sum_{k=1}^{N} \mathrm{pr}_k^2$*, then*

$$\lim_{n\to\infty} T_n(f) = f(\omega),$$

for every nonnegative function $f \in E \cap \mathcal{F}_b(\Omega)$*, which is continuous at* ω*. The conclusion occurs for all functions* $f \in E \cap \mathcal{F}_b(\Omega)$ *continuous at* ω *when the functionals* T_n *are weakly nonlinear and monotone increasing.*

The proof is similar to that of Theorem 5.4.1 and the details are left to the reader as an exercise.

5.5 The Case of Convergence in Measure

The convergence almost everywhere is not the only important mode of convergence for sequences of operators acting on spaces of measurable functions associated with a measure space (Ω, Σ, μ). Probability and Statistics highlight the case of *convergence in measure*, which technically means that

$$\lim_{n\to\infty} \mu\left(\{x \in \Omega : |f_n(x) - f(x)| \geq \varepsilon\}\right) = 0$$

for every $\varepsilon > 0$. When $\mu(\Omega) < \infty$, every sequence of measurable functions f_n that converges almost everywhere to a function f, also converges to f in measure. The converse assertion is false: there are sequences of measurable functions on $[0, 1]$ that converge to zero in Lebesgue measure but do not converge at any point. See Bogachev [25, Theorem 2.2.3, p. 111 and Example 2.2.4, p. 112].

The analog of Theorem 5.4.1 in the case of convergence in measure is as follows.

Theorem 5.5.1 *Suppose that* K *is a compact subset of the Euclidean space* $\mathbb{R}^N$ *endowed with a positive Borel measure* μ*, and let* $(T_n)_n$ *be a sequence of sublinear and monotone increasing operators from* $C(K)$ *into itself such that*

$$\lim_{n\to\infty} \mu\left(\{\mathbf{x} \in K : |T_n(f)(\mathbf{x}) - f(\mathbf{x})| \geq \varepsilon\}\right) = 0$$

for each $\varepsilon > 0$ *and each of the functions* $1, \pm \mathrm{pr}_1, \ldots, \pm \mathrm{pr}_N$ *and* $\sum_{k=1}^{N} \mathrm{pr}_k^2$*.*

Then this convergence occurs for all nonnegative functions $f \in C(K)$*. It occurs for all functions in* $C(K)$ *provided that the operators* T_n *are weakly nonlinear and monotone increasing.*

Proof Let $f \in C(K)$ be a nonnegative function, and let $\varepsilon > 0$ arbitrarily fixed. Due to the uniform continuity of the function f, for every $\alpha \in (0, \varepsilon/2)$, there is $\delta > 0$ such that

$$|f(\mathbf{t}) - f(\mathbf{x})| \leq \alpha + \delta \|\mathbf{t} - \mathbf{x}\|^2 \quad \text{for every } \mathbf{t}, \mathbf{x} \in K.$$

Proceeding as in the proof of Theorem 5.4.1, we infer that for all $\mathbf{x} \in K$ and $n \in \mathbb{N}$, we have

$$\begin{aligned}|T_n(f) - f(\mathbf{x})| &= |T_n(f) - T_n(f(\mathbf{x}) \cdot 1) + f(\mathbf{x})T_n(1) - f(\mathbf{x})| \\ &\leq T_n(|f - f(\mathbf{x}) \cdot 1|) + f(\mathbf{x})|T_n(1) - 1| \\ &\leq \alpha + \delta \left[T_n \left(\sum_{k=1}^{N} \mathrm{pr}_k^2 \right) + 2 \sum_{k=1}^{N} (M - \mathrm{pr}_k(\mathbf{x})) T_n \left(\mathrm{pr}_k\right)\right) \\ &\quad \left. + 2M \sum_{k=1}^{N} T_n \left(-\mathrm{pr}_k\right) + \|\mathbf{x}\|^2 T_n(1) \right] + f(\mathbf{x})|T_n(1) - 1|,\end{aligned}$$

where $M = \max \left\{ \mathrm{pr}_1(\mathbf{x}), \ldots, \mathrm{pr}_N(\mathbf{x}), 0 : \mathbf{x} \in K \right\}$.

Choose a rank N_0 such that $\|f\|_\infty |T_n(1) - 1| \leq \alpha$ for every $n \geq N_0$. Then for $n \geq N_0$, the set

$$\{\mathbf{x} \in K : |T_n(f)(\mathbf{x}) - f(\mathbf{x})| \geq \varepsilon\}$$

is included in the set of points $\mathbf{x} \in K$ where

$$\begin{aligned}T_n \left(\sum_{k=1}^{N} \mathrm{pr}_k^2 \right) (\mathbf{x}) &+ 2 \sum_{k=1}^{N} (M - \mathrm{pr}_k(\mathbf{x})) T_n \left(\mathrm{pr}_k\right)(\mathbf{x})) \\ &+ 2M \sum_{k=1}^{N} T_n \left(-\mathrm{pr}_k\right)(\mathbf{x}) + \|x\|^2 T_n(1)(\mathbf{x}) \geq \frac{\varepsilon - 2\alpha}{\delta}.\end{aligned}$$

This implies that

$$\begin{aligned}&\mu(\{\mathbf{x} \in K : |T_n(f)(\mathbf{x}) - f(\mathbf{x})| \geq \varepsilon\}) \\ &\quad \leq \mu \left(\left\{ \mathbf{x} \in K : T_n(\textstyle\sum_{k=1}^{N} \mathrm{pr}_k^2)(\mathbf{x}) + \cdots \geq (\varepsilon - 2\alpha)/\delta \right\} \right)\end{aligned}$$

for every $n \geq N_0$. Taking into account our hypotheses and the fact that the sum of sequences convergent in measure is also a sequence convergent in measure (see Bogachev [25, Corollary 2.2.6, p. 113]) we conclude the proof of the first part of Theorem 5.5.1.

For the second part, if $f \in C(K)$ is an arbitrary real-valued function we will apply the preceding reasoning to $f + \|f\|_\infty \geq 0$ to infer that $T_n(f + \|f\|_\infty) \to f + \|f\|_\infty$ in measure. We have

$$T_n(f + \|f\|_\infty) = T_n(f) + \|f\|_\infty T_n(1)$$

because the operators T_n are assumed to be weakly nonlinear and $T_n(1) \to 1$ by our hypotheses. The proof ends by using again the algebraic operations with sequences convergent in measure. ■

When K is only a locally compact subset of the Euclidean space $\mathbb{R}^N$, then Theorem 5.5.1 holds for every sequence of weakly nonlinear and monotone increasing operators $T_n : C_b(K) \to C_b(K)$, the conclusion being that

$$\lim_{n\to\infty} \mu\left(\{\mathbf{x} \in H : |T_n f)(\mathbf{x}) - f(\mathbf{x})| \geq \varepsilon\}\right) = 0$$

for every $\varepsilon > 0$ and every compact subset H of K.

5.6 The Case of Convergence in L^p-Norm

The existing literature includes many papers containing various generalizations of Korovkin's theorem in the context of operators acting on L^p-spaces. See Altomare [7], Altomare-Campiti [12], Berens-DeVore [21, 22], Donner [54, 55], Swetits-Wood [175], and Wulbert [188], to cite just a few. However, all of them refer to the case of linear and positive operators. The next theorem provides a nonlinear generalization based on sequences of weakly nonlinear and monotone operators.

Theorem 5.6.1 *Let μ be a positive Borel measure on $\mathbb{R}^N$ with compact support and let $(T_n)_n$ be a sequence of sublinear and monotone operators from the Banach lattice $L^p(\mu)$ into itself, where $p \in [1, \infty)$. If $M = \sup_n \|T_n\| < \infty$ and*

$$T_n(f) \to f \quad \text{in } L^p(\mu),$$

for each of the test functions $1, \ \pm \mathrm{pr}_1, \ldots, \ \pm \mathrm{pr}_N$ and $\sum_{k=1}^N \mathrm{pr}_k^2$, then this convergence occurs for all nonnegative functions $f \in L^p(\mu)$. It occurs for all functions in $L^p(\mu)$ provided that the operators T_n are weakly nonlinear and monotone.

Proof Let $f \in L^p(\mu)$, $f \geq 0$. Since $C_c(\mathbb{R}^N)$ is dense into $L^p(\mu)$ in the L^p-norm (see [25, Corollary 4.2.2, p. 252]) and the lattice operations in a Banach lattice are continuous [129, Proposition 1.1.6, p. 6], it follows that for every $\varepsilon > 0$, there exists a nonnegative continuous function g with compact support $\operatorname{supp} g$ such that

$$\|f - g\|_{L^p} < \varepsilon.$$

Then

$$\begin{aligned}
\|T_n(f) - f\|_{L^p} &\leq \|T_n(f) - T_n(g)\|_{L^p} + \|T_n(g) - gT_n(1)\|_{L^p} \\
&\quad + \|gT_n(1) - g\|_{L^p} + \|g - f\|_{L^p} \\
&\leq \|T_n\| \, \|f - g\|_{L^p} + \|T_n(g) - gT_n(1)\|_{L^p} + \|g\|_\infty \, \|T_n(1) - 1\|_{L^p} + \|f - g\|_{L^p} \\
&\leq \varepsilon(1 + M) + \|T_n(|g - g(x)|)(x)\|_{L^p} + \|g\|_\infty \, \|T_n(1) - 1\|_{L^p} .
\end{aligned}$$

Using the uniform continuity of g, one can infer (by reductio ad absurdum) that there exists a number $\delta \geq \|g\|_\infty + 1$ such that

$$|g(y) - g(x)| \leq \varepsilon + \delta \cdot \|y - x\|^2$$

for all x and y in an open and bounded neighborhood of the support of g, say $\{z : d(z, \operatorname{supp} g) < 1\}$ and therefore for all x and y in $\mathbb{R}^N$.

According to our hypotheses, there exists a rank n_0 such that

$$\|T_n(h) - h\|_{L^p} < \varepsilon/\delta$$

for all $n \geq n_0$ and all $h \in \{1, \pm \mathrm{pr}_1, \ldots, \pm \mathrm{pr}_N, \sum_{k=1}^N \mathrm{pr}_k^2\}$.

Put

$$\alpha = \max_{x \in K} \left\{ \left|\mathrm{pr}_1(x)\right|, \ldots, \left|\mathrm{pr}_N(x)\right| \right\},$$

where K denotes the support of μ. We have

$$\begin{aligned}
\|T_n(|g - g(x)|)(x)\|_{L^p} &\leq \|T_n(\varepsilon)(x) + \delta T_n(\|y - x\|^2)(x)\|_{L^p} \\
&\leq \varepsilon\|T_n(1)\|_{L^p} + \delta\|T_n(\|y - x\|^2)(x)\|_{L^p} \\
&\leq \varepsilon(\|T_n(1) - 1\|_{L^p} + \mu\,(K)^{1/p}) + \delta\|T_n(\|y - x\|^2)(x)\|_{L^p} \\
&\leq \varepsilon\left(\varepsilon + \mu(K)^{1/p}\right) + \delta\|T_n(\|y - x\|^2)(x)\|_{L^p}.
\end{aligned}$$

Since

$$\|y - x\|^2 = \sum_{k=1}^{N} \mathrm{pr}_k^2(y) + 2\sum_{k=1}^{N}(\alpha - \mathrm{pr}_k(x))\,\mathrm{pr}_k(y) + \sum_{k=1}^{N} \mathrm{pr}_k^2(x) + 2\alpha\sum_{k=1}^{N}\left(-\mathrm{pr}_k(y)\right),$$

it follows that

$$\begin{aligned}
T_n(\|y - x\|^2)(x) \le{}& T_n(\sum_{k=1}^{N} \mathrm{pr}_k^2)(x) + 2\sum_{k=1}^{N}(\alpha - \mathrm{pr}_k(x))T_n(\mathrm{pr}_k)(x)\\
&+ 2\alpha\sum_{k=1}^{N} T_n(-\mathrm{pr}_k)(x) + \sum_{k=1}^{N} \mathrm{pr}_k^2(x)T_n(1)(x)\\
={}& T_n(\sum_{k=1}^{N} \mathrm{pr}_k^2)(x) - \sum_{k=1}^{N} \mathrm{pr}_k^2(x) + 2\sum_{k=1}^{N}(\alpha - \mathrm{pr}_k(x))[T_n(\mathrm{pr}_k)(x) + T_n(-\mathrm{pr}_k)(x)]\\
&+ 2\sum_{k=1}^{N} \mathrm{pr}_k(x)[T_n(-\mathrm{pr}_k)(x) + \mathrm{pr}_k(x)] - \sum_{k=1}^{N} \mathrm{pr}_k^2(x) + \sum_{k=1}^{N} \mathrm{pr}_k^2(x)T_n(1)(x),
\end{aligned}$$

whence

$$\begin{aligned}
\delta \left\| T_n(\|y - x\|^2(x) \right\|_{L^p} \le{}& \delta \| T_n(\sum_{k=1}^{N} \mathrm{pr}_k^2)(x) - \sum_{k=1}^{N} \mathrm{pr}_k^2(x)\|_{L^p}\\
&+ 4\alpha\delta\sum_{k=1}^{N} \|T_n(-\mathrm{pr}_k)(x) + \mathrm{pr}_k(x)\|_{L^p} + 4\alpha\delta\sum_{k=1}^{N} ||T_n(\mathrm{pr}_k)(x) - \mathrm{pr}_k(x)\|_{L^p}\\
&+ 2\alpha\delta\sum_{k=1}^{N} \|T_n(-\mathrm{pr}_k)(x) + \mathrm{pr}_k(x)\|_p + \alpha^2\delta\sum_{k=1}^{N} \|T_n(1) - 1\|_{L^p}\\
&\le \varepsilon(1 + 10N\alpha + N\alpha^2)
\end{aligned}$$

for n sufficiently large. This ends the proof in the case where $f \ge 0$.

In the general case, notice first that $T_n(f) \to f$ in the L^p-norm for every $f \in L^p(\mu)$ with $\|f\|_\infty < \infty$. Indeed, $T_n(f + \|f\|_\infty) \to f + \|f\|_\infty$ by the discussion above and $T_n(f + \|f\|_\infty) = T_n(f) + \|f\|_\infty T_n(1)$ due to the fact that the operators T_n are weakly nonlinear.

If f is a real-valued function in $L^p(\mu)$, then it can be approximated by functions $h \in C_c(\mathbb{R}^N)$. Every such function h is bounded in the sup-norm, so that $\|T_n(h) - h\|_{L^p} \to 0$. Finally, the inequality

$$\|T_n(f) - f\| \leq \|T_n(f) - T_n(h)\|_{L^p} + \|T_n(h) - h\|_{L^p} + \|h - f\|_{L^p}$$
$$\leq M\, \|f - h\|_{L^p} + \|T_n(h) - h\|_{L^p} + \|h - f\|_{L^p},$$

allows us to conclude that $\|T_n(f) - f\|_{L^p} \to 0$. ■

It is conceivable that Theorem 5.6.1 still works in the case of an arbitrary finite Borel measure on $\mathbb{R}^N$, but at the moment this remains an open question.

In the setting of linear and continuous operators acting on $L^1[0, 1]$, Wulbert [188] has proved a Korovkin-type theorem that avoids the hypothesis of monotonicity. It is an open question whether his result admits an analog within the framework of continuous sublinear operators $T_n : L^1[0, 1] \to L^1[0, 1]$.

5.7 Operator Variants of Korovkin Theorem

All Korovkin-type theorems presented in the previous sections (except Altomare's Theorem 5.4.5) refer to the convergence of certain sequences of sublinear and monotone increasing operators to the identity operator. Wang [181], Guessab-Schmeisser [95] and Popa [153, 154] have noticed the possibility to prove (in the context of linear and positive operators) results concerning the convergence to a limit differing from the identity.

Recall here the results of Popa.

Theorem 5.7.1 (Popa [153]) *Suppose that K is a compact Hausdorff space, $A : C[a, b] \to C(K)$ and $T_n : C[a, b] \to C(K)$ $(n \in \mathbb{N})$ are positive linear operators such that*

$$A(1)A(e_2) = [A(e_1)]^2 \tag{5.16}$$

and

$$A(1)(t) > 0 \quad \text{for every } t \in K.$$

If

$$\lim_{n\to\infty} T_n(1) = A(1), \quad \lim_{n\to\infty} T_n(e_1) = A(e_1) \text{ and } \lim_{n\to\infty} T_n(e^2) = A(e^2),$$

then $\lim_{n\to\infty} T_n(f) = A(f)$ *for every* $f \in C[a, b]$.

Here, e_1 and e_2 stand for the functions defined on $[a, b]$ by the formulas $e_1(t) = t$ and $e_2(t) = t^2$.

Theorem 5.7.2 (Popa [154]) *Consider the positive linear operators L_n and U, from $C_{2\pi}(\mathbb{R})$ into itself, which satisfy the operator functional equation*

$$[U(\sin)]^2 + [U(\cos)]^2 = [U(1)]^2 \tag{5.17}$$

and the condition

$$U(1)(t) > 0 \quad \text{for every } t \in \mathbb{R}.$$

If

$$\lim_{n\to\infty} L_n(1) = U(1), \quad \lim_{n\to\infty} L_n(\cos) = U(\cos) \text{ and } \lim_{n\to\infty} L_n(\sin) = U(\sin),$$

then $\lim_{n\to\infty} L_n(f) = U(f)$, *for every* $f \in C_{2\pi}(\mathbb{R})$.

A characterization of the solutions of the operator functional equations (5.16) and (5.17) was provided by Altomare, who proved that they are necessarily weighted composition operators. See [11, Corollary 2.4 and Corollary 3.2].

The extension of Popa's Theorem 5.7.1 to our nonlinear context is based on the same basic ingredient, an a priori estimate, stated below as Lemma 5.7.3. Naturally, the details of the proof are a considerably more delicate.

The notation

$$u = (1, \ldots, 1)$$

will designate the natural order unit of the Banach lattice $\mathbb{R}^N$. Every compact subset K of $\mathbb{R}^N$ can be moved into the positive cone $\mathbb{R}^N_+$ via a translation of the form $T_\alpha : x \to x+\alpha u$, associated with a number $\alpha \geq 0$. The smallest α doing this job will be denoted $\alpha(K, u)$, and we will refer to it as the *deficit of positivity* of K in the direction u. We have

$$\alpha(K, u) = \inf\left\{\alpha \geq 0 : K + \alpha u \subset \mathbb{R}^N_+\right\}$$

and $\alpha(K, u) = 0$ if $K \subset \mathbb{R}^N_+$.

Lemma 5.7.3 *Suppose that K is a compact subset of the Euclidean space $\mathbb{R}^N$, X is a compact Hausdorff space, and T and A are weakly nonlinear and monotone increasing operators from $C(K)$ into $C(X)$. Then for every function $f \in C(K)$, and every $\varepsilon > 0$, there exists $\delta > 0$ such that*

$$
\begin{aligned}
&|T(f)A(1) - T(1)A(f)| \\
&\quad \leq \varepsilon T(1)A(1) + \delta \Bigg\{ A(1)[T(\sum_{k=1}^{N} (\mathrm{pr}_k + \alpha)^2) - A(\sum_{k=1}^{N} (\mathrm{pr}_k + \alpha)^2)] \\
&\qquad + A(\sum_{k=1}^{N} (\mathrm{pr}_k + \alpha)^2)[T(1) - A(1)] \\
&\qquad + 2[A(1)A(\sum_{k=1}^{N} (\mathrm{pr}_k + \alpha)^2) - \sum_{k=1}^{N} (A(-\mathrm{pr}_k - \alpha))^2] \\
&\qquad + 2\sum_{k=1}^{N} A(-\mathrm{pr}_k - \alpha)[A(-\mathrm{pr}_k - \alpha) - T(-\mathrm{pr}_k - \alpha)] \Bigg\},
\end{aligned}
$$

whenever $\alpha \geq \alpha(K, u)$.

Proof Let $f \in C(K)$ and $\varepsilon > 0$. According to Lemma 5.2.1, there is $\delta > 0$ such that for all $x, y \in K$ we have

$$
\begin{aligned}
|f(x) - f(y)| &\leq \varepsilon + \delta \, \|x - y\|^2 \\
&= \varepsilon + \delta \, \|x + \alpha u - (y + \alpha u)\|^2 .
\end{aligned}
$$

Here α is an arbitrarily fixed number such that $\alpha \geq \alpha(K, u)$. Viewing y as a parameter, the above inequality can be rewritten as

$$
|f - f(y)| \leq \varepsilon + \delta \left[\sum_{k=1}^{N} (\mathrm{pr}_k + \alpha)^2 - 2\sum_{k=1}^{N} (\mathrm{pr}_k(y) + \alpha)(\mathrm{pr}_k + \alpha) + \sum_{k=1}^{N} (\mathrm{pr}_k(y) + \alpha)^2 \right].
$$

Our choice of α makes $\mathrm{pr}_k + \alpha \geq 0$ for every k and this will enable us to value the special properties of the operator T.

Suppose for a moment that f is nonnegative. Then, according to Remark 4.1.14,

$$
\begin{aligned}
&|T(f) - f(y)T(1)| \leq T(|f - f(y) \cdot 1|) \\
&\quad \leq \varepsilon T(1) + \delta \Bigg[T(\sum_{k=1}^{N} (\mathrm{pr}_k + \alpha)^2) + 2\sum_{k=1}^{N} (\mathrm{pr}_k(y) + \alpha)\, T(-\mathrm{pr}_k - \alpha) \\
&\qquad + \sum_{k=1}^{N} (\mathrm{pr}_k(y) + \alpha)^2\, T(1) \Bigg],
\end{aligned}
$$

which yields, for each $t \in X$, the following inequality in $C(X)$:

$$
\begin{aligned}
&|(T(f))(t) - (T(1))(t) \cdot f| \\
&\quad \le \varepsilon (T(1))(t) + \delta \left[T(\sum_{k=1}^{N} (\mathrm{pr}_k + \alpha)^2)(t) + 2 \sum_{k=1}^{N} (T(-\mathrm{pr}_k - \alpha))(t)(\mathrm{pr}_k + \alpha) \right. \\
&\qquad \left. + (T(1))(t) \sum_{k=1}^{N} (\mathrm{pr}_k + \alpha)^2 \right].
\end{aligned}
$$

Applying the weakly nonlinear and monotone increasing operator A to the both sides of the last inequality and taking into account that $-2T(-\mathrm{pr}_k - \alpha) \ge 0$, we obtain

$$
\begin{aligned}
&|(T(f))(t)A(1) - (T(1))(t)A(f)| \le A(|(T(f))(t) - (T(1))(t) \cdot f|) \\
&\qquad \le \varepsilon (T(1))(t)A(1) + \delta \left[T(\sum_{k=1}^{N} (\mathrm{pr}_k + \alpha)^2)(t)A(1) \right. \\
&\quad \left. - 2 \sum_{k=1}^{N} T(-\mathrm{pr}_k - \alpha)(t)A(-\mathrm{pr}_k - \alpha) + (T(1))(t)A(\sum_{k=1}^{N} (\mathrm{pr}_k + \alpha)^2) \right],
\end{aligned}
$$

whence

$$
\begin{aligned}
&|(T(f))(t)A(1)(t) - (T(1))(t)A(f)(t)| \\
&\le \varepsilon (T(1))(t)A(1)(t) + \delta \left\{ [(T(\sum_{k=1}^{N} (\mathrm{pr}_k + \alpha)^2)(t) - A(\sum_{k=1}^{N} (\mathrm{pr}_k + \alpha)^2)(t)]A(1)(t) \right. \\
&\qquad + (A(\sum_{k=1}^{N} (\mathrm{pr}_k + \alpha)^2)(t))\,[T(1)(t) - A(1)(t)] \\
&\quad + 2[A(1)(t)A(\sum_{k=1}^{N} (\mathrm{pr}_k + \alpha)^2)(t) - \sum_{k=1}^{N} (A(-\mathrm{pr}_k - \alpha))^2(t)] \\
&\qquad \left. + 2 \sum_{k=1}^{N} A(-\mathrm{pr}_k - \alpha)(t)\left[A(-\mathrm{pr}_k - \alpha)(t) - T(-\mathrm{pr}_k - \alpha)(t)\right] \right\}.
\end{aligned}
$$

As $t \in X$ was arbitrarily chosen, this ends the proof in the case of nonnegative functions.

The case where f is a real-valued function can be reduced to this one by replacing f by $f + \|f\|_\infty$. Indeed, using the property of weak nonlinearity of the operators T and A, we have

$$\begin{aligned}&T(f+\|f\|_{\infty})A(1)-A(f+\|f\|_{\infty})T(1)\\&=\left(T(f)+\|f\|_{\infty}T(1)\right)A(1)-\left(A(f)+\|f\|_{\infty}A(1)\right)T(1)\\&=T(f)A(1)-A(f)T(1),\end{aligned}$$

and the proof is complete. ∎

Corollary 5.7.4 *Under the hypotheses of Lemma 5.7.3, the following inequality holds:*

$$\begin{aligned}&\|T(f)A(1)-T(1)A(f)\|\\&\leq \varepsilon\|T(1)\|\cdot\|A(1)\|+\delta\Bigg\{[\|A(1)\|\cdot\|T(\sum_{k=1}^{N}(\mathrm{pr}_k+\alpha)^2)-A(\sum_{k=1}^{N}(\mathrm{pr}_k+\alpha)^2)\|\\&\qquad+\|A(\sum_{k=1}^{N}(\mathrm{pr}_k+\alpha)^2)\|\cdot\|T(1)-A(1)\|\\&\qquad+2\|A(1)A(\sum_{k=1}^{N}(\mathrm{pr}_k+\alpha)^2)-\sum_{k=1}^{N}(A(-\mathrm{pr}_k-\alpha))^2\|\\&\qquad+2\sum_{k=1}^{N}\|A(-\mathrm{pr}_k-\alpha)\|\cdot\|T(-\mathrm{pr}_k-\alpha)-A(-\mathrm{pr}_k-\alpha)\|\Bigg\}.\end{aligned}$$

We are now in a position to settle the case of uniform convergence.

Theorem 5.7.5 *Suppose that K is a compact subset of the Euclidean space $\mathbb{R}^N$, X is a compact Hausdorff space, and T_n $(n\in\mathbb{N})$ and A are weakly nonlinear and monotone operators from $E=C(K)$ into $C(X)$ such that*

$$A(1)\text{ is a strictly positive function}\tag{5.18}$$

and

$$A(1)A(\sum_{k=1}^{N}(\mathrm{pr}_k+\alpha)^2)=\sum_{k=1}^{N}(A(-\mathrm{pr}_k-\alpha))^2\tag{5.19}$$

for a suitable $\alpha\geq\alpha(K,u)$. Then

$$\lim_{n\to\infty}\|T_n(f)-A(f)\|=0\quad\text{for every } f\in C(K)$$

if and only if this property occurs for each of the functions

$$1, -(\mathrm{pr}_1 + \alpha), \ldots, -(\mathrm{pr}_N + \alpha) \text{ and } \sum\nolimits_{k=1}^{N} (\mathrm{pr}_k + \alpha)^2. \tag{5.20}$$

Moreover, if $K \subset \mathbb{R}_+^N$, then Theorem 5.7.5 works for $\alpha = 0$.

Proof The necessity part is clear. For sufficiency, notice first that we may restrict ourselves to the case where $K \subset \mathbb{R}_+^N$ by performing, if necessary, a change of variable of the form $y = x + \alpha u$. This allows us to continue the proof with $\alpha = 0$.

According to Corollary 5.7.4, for all $f \in C(K)$ and $\varepsilon > 0$, there exists $\delta > 0$ such that for all $n \in \mathbb{N}$, we have

$$\begin{aligned}
&\|T_n(f)A(1) - T_n(1)A(f)\| \\
&\quad \leq \varepsilon \|T_n(1)\| \cdot \|A(1)\| + \delta \Bigg\{ [\|A(1)\| \cdot \|T_n(\sum_{k=1}^{N} \mathrm{pr}_k^2) - A(\sum_{k=1}^{N} \mathrm{pr}_k^2)\| \\
&+ \|A(\sum_{k=1}^{N} \mathrm{pr}_k^2)\| \cdot \|T_n(1) - A(1)\| + 2\|A(1)A(\sum_{k=1}^{N} \mathrm{pr}_k^2) - \sum_{k=1}^{N} [A(-\mathrm{pr}_k)]^2\| \\
&\qquad + 2\sum_{k=1}^{N} \|A(-\mathrm{pr}_k)\| \cdot \|T_n(-\mathrm{pr}_k) - A(-\mathrm{pr}_k)\| \Bigg\},
\end{aligned}$$

whence, by using hypothesis (5.19), we infer that

$$\begin{aligned}
&\|T_n(f)A(1) - T_n(1)A(f)\| \\
&\quad \leq \varepsilon \|T_n(1)\| \cdot \|A(1)\| + \delta \Bigg\{ [\|A(1)\| \cdot \|T_n(\sum_{k=1}^{N} \mathrm{pr}_k^2) - A(\sum_{k=1}^{N} \mathrm{pr}_k^2)\| \\
&+ \|A(\sum_{k=1}^{N} \mathrm{pr}_k^2)\| \cdot \|T_n(1) - A(1)\| + 2\sum_{k=1}^{N} \|A(-\mathrm{pr}_k)\| \cdot \|T_n(-\mathrm{pr}_k) - A(-\mathrm{pr}_k)\| \Bigg\}.
\end{aligned}$$

Since $\lim_{n\to\infty} T_n(f) = A(f)$ for $f \in \left\{1, -\mathrm{pr}_1, \ldots, -\mathrm{pr}_N, \sum_{k=1}^{N} \mathrm{pr}_k^2\right\}$, the last inequality implies that

$$\lim_{n\to\infty} \|T_n(f)A(1) - T_n(1)A(f)\| = 0.$$

Combining this fact with the inequality

$$\|T_n(f)A(1) - A(1)A(f)\| \leq \|T_n(f)A(1) - T_n(1)A(f)\|$$

$$+ \|T_n(1)A(f) - A(1)A(f)\|$$

$$\leq \|T_n(f)A(1) - T_n(1)A(f)\| + \|A(f)\| \cdot \|T_n(1) - A(1)\|,$$

we infer that

$$\lim_{n\to\infty} \|[T_n(f) - A(f)]A(1)\| = 0.$$

By our hypotheses, the function $A(1)$ is strictly positive, so that $\inf_{t\in X} A(1)(t)$ is strictly positive. Since

$$[\inf_{t\in X} A(1)(t)] \cdot \|T_n(f) - A(f)\| \leq \|[T_n(f) - A(f)]A(1)\|,$$

it follows that

$$\lim_{n\to\infty} \|T_n(f) - A(f)\| = 0$$

and the proof is done. ■

In the particular case where A is an algebra homomorphism preserving the unit (in particular, if $A = \mathrm{I}_{C(K)}$), the conditions (5.19) and (5.18) are automatically fulfilled and Theorem 5.7.5 is covered by our previous results.

The condition $\lim_{n\to\infty} \|T_n(1) - A(1)\| = 0$ in Theorem 5.7.5 implies the equicontinuity of the operators T_n since $\|T_n\| = T_n(1)$ for all n.

Example 5.7.6 Given a continuous function $\varphi : [0, 1] \to [0, 1]$, one associates with it the sequence of nonlinear operators $T_n : C([0, 1]) \to C([0, 1])$ defined by the formulas

$$T_n(f)(x) = \sum_{k=0}^{n} p_{n,k}(\varphi(x)) \sup_{[k/(n+1)\leq t\leq (k+1)/(n+1)]} f(t),$$

where

$$p_{n,k}(x) = \binom{n}{k} x^k (1-x)^{n-k}.$$

Clearly, these operators are sublinear, invariant to translation, monotone increasing, and unital. Moreover, the sequence $(T_n)_n$ is pointwise convergent to the linear, positive, and unital operator

$$A : C([0, 1]) \to C([0, 1]), \quad A(f) = f \circ \varphi.$$

The details are covered by Theorem 5.7.5. Indeed, the condition (5.18) is trivial since A is unital. The fulfillment of condition (5.19) is a consequence of the fact that A is an algebra homomorphism preserving the unit; see the comments after the proof of Theorem 5.7.5. Using the combinatorial identities from the introduction to this chapter, one can show that $T_n(f) \to A(f)$ uniformly for each of the functions e_0, $-e_1$ and e_2; this set of test functions suffices since $[0, 1] \subset \mathbb{R}_+$. We have

$$\begin{aligned}
T_n(e_0)(x) &= 1 \\
T_n(-e_1)(x) &= \sum_{k=0}^{n} \binom{n}{k} (\varphi(x))^k (1-\varphi(x))^{n-k} \cdot (-k/(n+1)) \\
&= -\frac{n}{n+1} \sum_{k=0}^{n} \binom{n}{k} (\varphi(x))^k (1-\varphi(x))^{n-k} \frac{k}{n} = -\frac{n}{n+1}\varphi(x),
\end{aligned}$$

and

$$\begin{aligned}
T_n(e_2)(x) &= \sum_{k=0}^{n} \binom{n}{k} (\varphi(x))^k (1-\varphi(x))^{n-k} \cdot \frac{(k+1)^2}{(n+1)^2} \\
&= \left(\frac{n}{n+1}\right)^2 \cdot \sum_{k=0}^{n} \binom{n}{k} (\varphi(x))^k (1-\varphi(x))^{n-k} \frac{k^2}{n^2} \\
&\quad + 2\frac{n}{(n+1)^2} \sum_{k=0}^{n} \binom{n}{k} (\varphi(x))^k (1-\varphi(x))^{n-k} \frac{k}{n} + \frac{1}{(n+1)^2} \\
&= \left(\frac{n}{n+1}\right)^2 \left((\varphi(x))^2 + \frac{\varphi(x)(1-\varphi(x))}{n}\right) + \frac{2n}{(n+1)^2}\varphi(x) + \frac{1}{(n+1)^2},
\end{aligned}$$

Now it is clear that the operators T_n and A fulfill the hypotheses of Theorem 5.7.5, and we can conclude that $T_n(f) \to A(f)$, uniformly for all $f \in C([0, 1])$. A quantitative estimate of this convergence makes the objective of Example 5.8.6.

The result of Theorem 5.7.5 can be put in a more generality as follows:

Theorem 5.7.7 *Suppose that K is a compact subset of the Euclidean space $\mathbb{R}^N$, X is a compact Hausdorff space, and E is a Banach lattice that includes $C(K)$ in such a way that the canonical inclusion $i_K : C(K) \to E$ is a positive linear operator with a dense image. If T_n $(n \in \mathbb{N})$ and A are weakly nonlinear and monotone increasing operators from E into $C(X)$ such that*

$$A(1) \text{ is a strictly positive element,} \tag{5.21}$$

$$A(1)A(\sum_{k=1}^{N}(\mathrm{pr}_k+\alpha)^2)=\sum_{k=1}^{N}(A(-\mathrm{pr}_k-\alpha))^2 \textit{ for some } \alpha\geq d(K,u), \tag{5.22}$$

and

$$\sup_n \|T_n\|<\infty,$$

then $\lim_{n\to\infty}\|T_n(f)-A(f)\| = 0$ *for all* $f\in E$, *if and only if this property of convergence occurs for the following set of functions:*

$$1, -(\mathrm{pr}_1+\alpha), \ldots, -(\mathrm{pr}_N+\alpha) \textit{ and } \sum\nolimits_{k=1}^{N}(\mathrm{pr}_k+\alpha)^2. \tag{5.23}$$

In particular, Theorem 5.7.7 holds for the Banach lattices $E = L^p(\mu)$, where μ is a positive Borel measure on K. The fact that $L^p(\mu)$ includes $C(K)$ as a dense subspace can be covered from many sources, e.g., Bogachev [25, Corollary 4.2.2, p. 252].

Proof Let $f\in E$ and $\varepsilon>0$. Since the vector lattice $C(K)$ is dense in E , there exists $g\in C(K)$ such that

$$\|f-g\|<\varepsilon. \tag{5.24}$$

Since g is uniformly continuous, there exists $\delta>0$ such that

$$|g(y)-g(x)|\leq\varepsilon+\delta\cdot\|y-x\|^2$$

for all x and y in K.

According to Corollary 5.7.4, applied to $T_n\circ i_K$ and $A\circ i_K$, we have

$$\begin{aligned}
&\|T_n(g)A(1)-T_n(1)A(g)\|\\
&\leq\varepsilon\|T_n(1)\|\cdot\|A(1)\|+\delta\Bigg\{[\|A(1)\|\cdot\|T_n(\sum_{k=1}^{N}(\mathrm{pr}_k+\alpha)^2)-A(\sum_{k=1}^{N}(\mathrm{pr}_k+\alpha)^2)\|\\
&\qquad+\|A(\sum_{k=1}^{N}(\mathrm{pr}_k+\alpha)^2)\|\cdot\|T_n(1)-A(1)\|\\
&\qquad+2\|A(1)A(\sum_{k=1}^{N}(\mathrm{pr}_k+\alpha)^2)-\sum_{k=1}^{N}(A(-\mathrm{pr}_k-\alpha))^2\|\\
&\qquad+2\sum_{k=1}^{N}\|A(-\mathrm{pr}_k-\alpha)\|\cdot\|T_n(-\mathrm{pr}_k-\alpha)-A(-\mathrm{pr}_k-\alpha)\|\Bigg\}
\end{aligned}$$

whence, by denoting $M = \sup \|T_n(1)\|$ and taking into account the hypothesis (5.22), we obtain the inequality

$$\begin{aligned}
&\|T_n(g)A(1) - T_n(1)A(g)\| \\
&\quad \leq \varepsilon M \|A(1)\| + \delta \Bigg\{ [\|A(1)\| \cdot \|T_n(\sum_{k=1}^{N} (\mathrm{pr}_k + \alpha)^2) - A(\sum_{k=1}^{N} (\mathrm{pr}_k + \alpha)^2)\| \\
&\qquad + \|A(\sum_{k=1}^{N} (\mathrm{pr}_k + \alpha)^2)\| \cdot \|T_n(1) - A(1)\| \\
&\qquad + 2\sum_{k=1}^{N} \|A(-\mathrm{pr}_k - \alpha)\| \cdot \|T_n(-\mathrm{pr}_k - \alpha) - A(-\mathrm{pr}_k - \alpha)\| \Bigg\}.
\end{aligned}$$

Thus, under the presence of condition (5.23), we have

$$\lim_{n\to\infty} \|T_n(g)A(1) - T_n(1)A(g)\| = 0.$$

Since

$$\begin{aligned}
\|[T_n(g) - A(g)]A(1)\| &\leq \|T_n(g)A(1) - T_n(1)A(g) + T_n(1)A(g) - A(1)A(g)\| \\
&\leq \|T_n(g)A(1) - T_n(1)A(g)\| + \|A(g)\| \|T_n(1) - A(1)\|,
\end{aligned}$$

this yields

$$\lim_{n\to\infty} \|[T_n(g) - A(g)]A(1)\| = 0.$$

Proceeding as in the proof of Theorem 5.7.5, we infer that

$$\lim_{n\to\infty} \|T_n(g) - A(g)\| = 0$$

for all $g \in C(K)$. This property can be transferred to f due to the inequalities

$$\begin{aligned}
\|T_n(f) - A(f)\| &\leq \|T_n(f) - T_n(g)\| + \|T_n(g) - A(g)\| + \|A(g) - A(f)\| \\
&\leq \sup_n \|T_n\| \cdot \|f - g\| + \|T_n(g) - A(g)\| + \|A\| \cdot \|f - g\| \\
&\leq (\sup_n \|T_n\| + \|A\|)\varepsilon + \|T_n(g) - A(g)\|.
\end{aligned}$$

Therefore,

$$\lim_{n\to\infty} \|T_n(f) - A(f)\| = 0,$$

for all $f \in E$ and the proof is done. ■

When the property of translatability is replaced in Theorems 5.7.5 and 5.7.7 by translation invariance, the set of test functions can be simplified as follows:

$$1, -\mathrm{pr}_1, \ldots, -\mathrm{pr}_N \text{ and} \sum_{k=1}^{N} \left(\mathrm{pr}_k^2 + 2\alpha\, \mathrm{pr}_k\right).$$

As noticed by Korovkin [107, 108], his theorem mentioned in the introduction of this chapter also works when the unit interval is replaced by the unit circle

$$S^1 = \{(\cos\theta, \sin\theta) : \theta \in [0, 2\pi)\} \subset \mathbb{R}^2.$$

Since the cosine function represents the restriction of pr_1 to S^1 and the sine function represents the restriction of pr_2 to S^1, one can easily deduce from Theorem 5.7.5 the following more general result.

Theorem 5.7.8 *Suppose that T_n $(n \in \mathbb{N})$ and A are weakly nonlinear and monotone operators from $E = C_{2\pi}(\mathbb{R})$ into itself such that $A(1) = 1$ and*

$$A(3 + 2\cos + 2\sin) = (A(-1 - \cos))^2 + (A(-1 - \sin))^2. \tag{5.25}$$

Then

$$T_n(f) \to A(f) \quad \textit{uniformly for all } f \in C_{2\pi}(\mathbb{R})$$

if and only if this property occurs for each of the following test functions,

$$1, \quad -1 - \cos, \quad -1 - \sin \textit{ and } 3 + 2\cos + 2\sin.$$

Theorem 5.7.8 constitutes a nonlinear extension of Popa's Theorem 5.7.2.

Remark 5.7.9 Notice that the hypothesis (5.25) is automatically fulfilled when A is an algebra homomorphism preserving the unit (in particular, if $A = \mathrm{I}$). When the property of translatability is strengthened to invariance to translation, then the set of test functions in Theorem 5.7.8 can be replaced by

$$1, \quad -\cos, \quad -\sin \text{ and } 2\cos + 2\sin.$$

The analogs of Theorem 5.7.8 for some other cases of interest such as the 2-dimensional torus $S^1 \times S^1$ and the 2-dimensional sphere S^2 are left to the reader as an exercise.

An open problem that might be of interest in connection with Theorems 5.7.7 and 5.7.8 is as follows.

Problem 5.7.10 How large is the set of solutions of the functional equation

$$A(1)A\left(\sum_{k=1}^{N}(pr_k+\alpha)^2\right)=\sum_{k=1}^{N}(A(-pr_k-\alpha))^2,$$

among all weakly nonlinear and monotone increasing operators A (from $C(K)$ into $C(X)$)?

Also, consider the same question concerning the functional equation appearing in the statement of Theorem 5.7.8.

5.8 Quantitative Estimates Relative to the Uniform Convergence

In a very influential paper, Shisha and Mond [168] have put the classical theorem of Korovkin in a quantitative form, expressing the rate of convergence of $T_n(f)$ to f, in terms of the rates of convergence for the test functions 1, x and x^2. A similar result was noticed by Kiran Kumar and Vinaya [106] in the context of Popa's operator version of the Korovkin theorem. Its extension to the framework of weakly nonlinear and monotone increasing operators acting on spaces of continuous functions is due to the present authors. See [80].

Before entering the details, let us recall here a basic inequality concerning the *modulus of continuity* of a real-valued continuous function f defined on a compact subset K of $\mathbb{R}^N$,

$$\omega(f,\delta)=\sup\{|f(x)-f(y)| : x,y\in K,\ \|x-y\|\leq\delta\},\qquad \delta>0.$$

Lemma 5.8.1 (Shisha and Mond [168], p. 1197) *For all $x, y \in K$ and $\delta > 0$, we have*

$$|f(x)-f(y)|\leq(1+\|x-y\|^2\delta^{-2})\omega(f,\delta).$$

Proof Indeed, if $\|x-y\|\leq\delta$, then

$$|f(x)-f(y)|\leq\omega(f,\|x-y\|)\leq\omega(f,\delta)\leq(1+\|x-y\|^2\delta^{-2})\omega(f,\delta),$$

while if $\|x - y\| > \delta$, we obtain, by taking into account the following well-known property of the modulus of continuity,

$$\omega(f, \lambda\delta) \leq (1 + \lambda)\omega(f, \delta) \quad \text{for } \lambda, \delta > 0,$$

the inequality

$$|f(x) - f(y)| \leq \omega(f, \|x - y\|) \leq (1 + \|x - y\|^2\delta^{-2})\omega(f, \delta).$$

The proof is complete. ∎

One can now formulate the following quantitative version of Theorem 5.7.5.

Theorem 5.8.2 *Let K be a compact subset of the Euclidean orthant $\mathbb{R}_+^N$, X a compact Hausdorff space, and T_n ($n \in \mathbb{N}$) and A weakly nonlinear and monotone increasing operators from $E = C(K)$ into $C(X)$, such that $A(1)$ is a strictly positive function. Then,*

$$\|T_n(f) - A(f)\| \leq M\{\|T_n(1) - A(1)\| \cdot \|A(f)\| \\ +(\|T_n(1)A(1)\| + 1)\omega(f, \mu_n)\}$$

for all $f \in C(K)$ and $n \in \mathbb{N}$, where

$$M = 1/\inf_{x\in X}(A(1)(x))$$

and μ_n denotes the square root of

$$\left\| T_n\left(\sum_{k=1}^N \mathrm{pr}_k^2\right) A(1) - 2\sum_{k=1}^N A\left(-\mathrm{pr}_k\right) T_n\left(-\mathrm{pr}_k\right) + A\left(\sum_{k=1}^N \mathrm{pr}_k^2\right) T_n(1) \right\|.$$

In the particular case when $T_n(1) = A(1) = 1$ for all n, the previous estimate reduces to

$$\|T_n(f) - A(f)\| \leq 2\omega(f, \mu_n).$$

Proof Since $A(1)(X)$ is a compact subset of $\mathbb{R}$ and $A(1)(x) > 0$ for all $x \in X$, we get $\inf_{x\in X} A(1)(x) > 0$. Also, for all $f \in C(K)$, it follows that

$$\frac{1}{M}\|T_n(f) - A(f)\| = \inf_{x\in X} A(1)(x) \cdot \sup_{x\in X} |T_n(f)(x) - A(f)(x)|$$

$$\leq \sup_{x \in X} |(T_n(f)(x) - A(f)(x))\, A(1)(x)|$$

$$= \|(T_n(f) - A(f))A(1)\|, \tag{5.26}$$

which reduces the proof to the inequality

$$\begin{aligned}
&\|(T_n(f) - A(f))A(1)\| \\
&\leq \{\|T_n(1) - A(1)\| \cdot \|A(f)\| + (\|T_n(1)A(1)\| + 1)\omega(f, \mu_n)\}.
\end{aligned} \tag{5.27}$$

For this last one, by Lemma 5.8.1, we get

$$|f(x) - f(y)| \leq (1 + \|x - y\|^2\delta^{-2})\omega(f, \delta),$$

whenever $x, y \in K$ and $\delta > 0$, a fact that yields (viewing y as a parameter) the inequality

$$|f - f(y)| \leq \left[1 + \delta^{-2}\left(\sum_{k=1}^{N} \mathrm{pr}_k^2 - 2\sum_{k=1}^{N} \mathrm{pr}_k(y)\, \mathrm{pr}_k + \sum_{k=1}^{N} \mathrm{pr}_k^2(y) \cdot 1\right)\right] \omega(f, \delta).$$

Suppose for a moment that $f \geq 0$. According to Remark 4.1.14, it follows that

$$\begin{aligned}
|T_n(f) - f(y)T_n(1)| &\leq T_n(|f - f(y) \cdot 1|) \\
&\leq \left[T_n(1) + \delta^{-2}\left(T_n(\sum_{k=1}^{N} (\mathrm{pr}_k)^2) + 2\sum_{k=1}^{N} (\mathrm{pr}_k(y))\, T_n\left(-\mathrm{pr}_k\right)\right.\right. \\
&\qquad \left.\left. + \sum_{k=1}^{N} (\mathrm{pr}_k(y))^2\, T_n(1)\right)\right] \omega(f, \delta),
\end{aligned}$$

which, for each $x \in X$, implies the following inequality in $C(X)$:

$$\begin{aligned}
&|\,(T_n(f))\,(x) - T_n(1)(x) \cdot f| \\
&\leq \left[T_n(1)(x) + \delta^{-2}\left(T_n(\sum_{k=1}^{N} \mathrm{pr}_k^2)(x) + 2\sum_{k=1}^{N} T_n\left(-\mathrm{pr}_k\right)(x)\ \mathrm{pr}_k\right.\right. \\
&\qquad \left.\left. + \sum_{k=1}^{N} T_n(1)(x)\, \mathrm{pr}_k^2\right)\right] \omega(f, \delta).
\end{aligned}$$

Applying now the weakly nonlinear and monotone operator A to both sides of this last inequality and observing that $-2T_n(-pr_k) \geq 0$, we obtain

$$|(T_n(f))(x)A(1)-(T_n(1))(x)A(f)| \le A\left(|(T_n(f))(x)-(T_n(1))(x)\cdot f|\right)$$
$$\le \left[T_n(1)(x)A(1)+\delta^{-2}\left(T_n(\sum_{k=1}^{N}\mathrm{pr}_k^2)(x)A(1)-2\sum_{k=1}^{N}A\left(-\mathrm{pr}_k\right)T_n\left(-\mathrm{pr}_k\right)(x)\right.\right.$$
$$\left.\left.+A(\sum_{k=1}^{N}\mathrm{pr}_k^2)T_n(1)(x)\right)\right]\omega(f,\delta),$$

whence

$$|(T_n(f))(x)A(1)(x)-(T_n(1))(x)A(f)(x)|$$
$$\le \left[T_n(1)(x)A(1)(x)+\delta^{-2}\left(T_n(\sum_{k=1}^{N}\mathrm{pr}_k^2)(x)A(1)(x)\right.\right.$$
$$\left.\left.-2\sum_{k=1}^{N}A\left(-\mathrm{pr}_k\right)(x)T_n(-\mathrm{pr}_k)(x)+A(\sum_{k=1}^{N}\mathrm{pr}_k^2)(x)T_n(1)(x)\right)\right]\omega(f,\delta),$$

for all $x\in X$. Taking the supremum over $x\in X$, we arrive at

$$\|(T_n(f))A(1)-(T_n(1))A(f)\| \tag{5.28}$$
$$\le \left[\|T_n(1)A(1)\|+\delta^{-2}\left(\left\|T_n(\sum_{k=1}^{N}\mathrm{pr}_k^2)A(1)-2\sum_{k=1}^{N}A\left(-\mathrm{pr}_k\right)T_n\left(-\mathrm{pr}_k\right)\right.\right.\right.$$
$$\left.\left.\left.+A(\sum_{k=1}^{N}\mathrm{pr}_k^2)T_n(1)\right\|\right)\right]\omega(f,\delta).$$

The case when $f\in C(K)$ is real-valued can be reduced to the precedent one by replacing f by $f+\|f\|_\infty$. Indeed, using the property of invariance to translation of the operators T_n and A, we have

$$\begin{aligned}&T_n(f+\|f\|_\infty)A(1)-A(f+\|f\|_\infty)T_n(1)\\&=\left(T_n(f)+\|f\|_\infty T_n(1)\right)A(1)-\left(A(f)+\|f\|_\infty A(1)\right)T_n(1)\\&=T_n(f)A(1)-A(f)T_n(1),\end{aligned}$$

which implies the validity of the inequality (5.28) for all functions $f\in C(K)$.

Supposing that $\mu_n\neq 0$, we deduce from (5.28), applied for $\delta=\mu_n$, the relation

$$\|(T_n(f))A(1)-(T_n(1))A(f)\|\le(\|T_n(1)A(1)\|+1)\omega(f,\mu_n).$$

Then

$$\begin{aligned}\|T_n(f)A(1) - A(1)A(f)\| &\leq \|T_n(f)A(1) - T_n(1)A(f)\| \\ &\quad + \|T_n(1)A(f) - A(1)A(f)\| \\ &\leq (\|T_n(1)A(1)\| + 1)\omega(f, \mu_n) + \|A(f)\| \cdot \|T_n(1) - A(1)\|\end{aligned}$$

which proves the inequality (5.27).

It remains to consider the case $\mu_n = 0$. Then the inequality (5.28) reduces to

$$\|T_n(f)A(1) - T_n(1)A(f)\| = 0$$

and we have

$$\begin{aligned}\|T_n(f)A(1) - A(1)A(f)\| &\leq \|T_n(f)A(1) - T_n(1)A(f)\| + \|T_n(1)A(f) - A(1)A(f)\| \\ &\leq \|A(f)\| \cdot \|T_n(1) - A(1)\|,\end{aligned}$$

which implies the inequality stated by Theorem 5.8.2 for $\mu_n = 0$. ■

Remark 5.8.3 Theorem 5.8.2 implies the qualitative result stated by Theorem 5.7.5. To prove this, notice that

$$\begin{aligned}T_n\left(\sum_{k=1}^N \mathrm{pr}_k^2\right) A(1) - 2\sum_{k=1}^N A\left(-\mathrm{pr}_k\right) T_n\left(-\mathrm{pr}_k\right) + A\left(\sum_{k=1}^N \mathrm{pr}_k^2\right) T_n(1) \\ = \left[T_n\left(\sum_{k=1}^N \mathrm{pr}_k^2\right) - A\left(\sum_{k=1}^N \mathrm{pr}_k^2\right)\right] A(1) \\ - 2\sum_{k=1}^N A(-\mathrm{pr}_k)[T_n(-\mathrm{pr}_k) - A(-\mathrm{pr}_k)] \\ + [T_n(1) - A(1)]A\left(\sum_{k=1}^N \mathrm{pr}_k^2\right) + 2\Delta,\end{aligned}$$

where

$$\Delta = A(1)A\left(\sum_{k=1}^N \mathrm{pr}_k^2\right) - \sum_{k=1}^N (A(-\mathrm{pr}_k))^2.$$

Consequently, if $\Delta = 0$ (which is one of the hypotheses of Theorem 5.7.5), it follows that $\mu_n \to 0$ whenever

$$\lim_{n\to\infty} T_n(1) = A(1),$$

$$\lim_{n\to\infty} T_n(-\operatorname{pr}_k) = A(-\operatorname{pr}_k), \quad k = 1, \ldots, N$$

and

$$\lim_{n\to\infty} T_n\left(\sum_{k=1}^{N} \operatorname{pr}_k^2\right) = A\left(\sum_{k=1}^{N} \operatorname{pr}_k^2\right),$$

uniformly on X. According to Theorem 5.8.2,

$$\lim_{n\to\infty} T_n(f) = A(f)$$

uniformly on X and the proof is done.

In the case of the functions $f \in C([0, 1])$, Theorem 5.8.2 reads as follows.

Corollary 5.8.4 *Let $T_n : C([0, 1]) \to C([0, 1])$ $(n \in \mathbb{N})$ be a sequence of weakly nonlinear and monotone increasing operators. Then for every weakly nonlinear and monotone increasing operator $A : C([0, 1]) \to C([0, 1])$ such that $A(1)(x) > 0$ for all $x \in [0, 1]$ and every function $f \in C([0, 1])$, we have the estimate*

$$\|T_n(f) - A(f)\| \le M\{\|T_n(1) - A(1)\| \cdot \|A(f)\| + (\|T_n(1)A(1)\| + 1)\omega(f, \mu_n)\},$$

where $M = (\inf_{x\in [0,1]} A(1)(x))^{-1}$ and

$$\mu_n = \|T_n(e_2)A(1) - 2T_n(-e_1)A(-e_1) + T_n(1)A(e_2)\|^{1/2}.$$

In addition, if $T_n(1) = A(1)$, then

$$\|T_n(f) - A(f)\| \le M(\|A(1)\|^2 + 1)\omega(f, \mu_n).$$

Corollary 5.8.4 extends Theorem 1 from Shisha-Mond [168] (proved by them in the particular case where all operators T_n are linear, continuous and positive and A is the identity operator on $C([0, 1])$).

In what follows we will illustrate Corollary 5.8.4 by some examples.

Example 5.8.5 Let $\varphi : [0, 1] \to [0, 1]$ be a continuous function. Associated with it is the sequence of Bernstein-type operators $\tilde{B}_n : C([0, 1]) \to C([0, 1])$, defined by the formulas

$$\tilde{B}_n(f)(x) = \sum_{k=0}^{n} \binom{n}{k} \varphi(x)^k (1 - \varphi(x))^{n-k} f(k/n),$$

and also the operator

$$A : C([0, 1]) \to C([0, 1]), \quad A(f) = f \circ \varphi.$$

These operators are linear, positive, and unital. Also, the operator A verifies the condition

$$A(1)A(e_2) = (A(e_1))^2.$$

Now, consider the operators $T_n : C([0, 1]) \to C([0, 1])$ defined by

$$T_n(f) = \max\left\{\tilde{B}_n(f), \tilde{B}_{n+1}(f)\right\}.$$

Obviously they are sublinear, monotone, and invariant to translation. Since $\tilde{B}_n(e_0)(x) = 1$, $\tilde{B}_n(-e_1)(x) = -\varphi(x)$ and

$$\tilde{B}_n(e_2)(x) = (\varphi(x))^2 + \frac{\varphi(x)(1 - \varphi(x))}{n},$$

it follows that $T_n(e_0)(x) = 1$, $T_n(-e_1)(x) = -\varphi(x)$ and

$$T_n(e_2)(x) = \max\{\tilde{B}_n(e_2)(x), \tilde{B}_{n+1}(e_2)(x)\} = (\varphi(x))^2 + \frac{\varphi(x)(1 - \varphi(x))}{n},$$

which by Corollary 5.8.4 implies that

$$\mu_n^2 = \left\| \frac{\varphi(1 - \varphi)}{n} \right\| \le \frac{1}{4n}.$$

In conclusion,

$$\|T_n(f) - A(f)\| \le 2\omega\left(f, \frac{1}{2\sqrt{n}}\right).$$

Example 5.8.6 Given the continuous function $\varphi : [0, 1] \to [0, 1]$, one associates with it the sequence of nonlinear operators $T_n : C([0, 1]) \to C([0, 1])$ defined by the formula

$$T_n(f)(x) = \sum_{k=0}^{n} \binom{n}{k} (\varphi(x))^k (1-\varphi(x))^{n-k} \sup_{[k/(n+1)\leq t\leq,(k+1)/(n+1)]} f(t)$$

and the composition operator A, which associates to each $f \in C([0,1])$ the function $f \circ \varphi$. According to Example 5.7.6, $T_n(f) \to A(f)$ uniformly for each $f \in C([0,1])$. Based on the calculations shown in that example, one obtains from Corollary 5.8.4 that

$$\begin{aligned}\mu_n^2 &= \left\| \left(\frac{n}{n+1}\right)^2 \left(\varphi^2 + \frac{\varphi(1-\varphi)}{n}\right) + \frac{2n\varphi}{(n+1)^2} + \frac{1}{(n+1)^2} - \frac{2n\varphi^2}{n+1} + \varphi^2 \right\| \\ &= \left\| \frac{3n\varphi + 1 - \varphi^2(n-1)}{(n+1)^2} \right\| \leq \frac{2}{n+1},\end{aligned}$$

for $n \geq 1$. In conclusion, for every $f \in C([0,1])$ and $n \geq 1$,

$$\|T_n(f) - A(f)\| \leq 2\omega\left(f, \frac{\sqrt{2}}{\sqrt{n+1}}\right).$$

Some information regarding the quantitative estimates related to convergence in L^p-norm can be found in the Comments section at the end of this chapter.

5.9 Feller's Probabilistic Scheme of Approximation

A well-known probabilistic scheme in constructing linear and positive operators of approximation is that due to Feller [63]. See also Altomare and Campiti [12], where Feller's main result is put in a more general form by their Theorem 5.2.2, pp. 286–288.

It is worth mentioning the equivalence between Korovkin's theorem and Feller's theorem, proved in [12, Theorem 5.2.3, pp. 289–290].

In what follows we will draw a parallel to Feller's technique of approximation, by using the more general framework of Choquet integral. This will generate a convenient way to construct weakly nonlinear and monotonic increasing approximation operators.

Let (Ω, Σ, μ) be a normalized capacity space. The *push-forward capacity* of μ through a measurable function $X : \Omega \to \mathbb{R}$ is the normalized capacity μ_X defined by the formula

$$\mu_X(B) = \mu(X^{-1}(B)) \quad \text{for every } B \in \mathcal{B}(\mathbb{R}).$$

Lemma 5.9.1 *A measurable function $f : \mathbb{R} \to \mathbb{R}$ is Choquet integrable with respect to the capacity μ_X if and only if the function $f \circ X$ is Choquet integrable with respect to μ, in which case,*

$$(C)\int_{\mathbb{R}} f\,\mathrm{d}\mu_X = (C)\int_{\Omega} f\circ X\,\mathrm{d}\mu.$$

More generally, for every $B \in \mathcal{B}(\mathbb{R})$, *we have*

$$(C)\int_{B} f\,\mathrm{d}\mu_X = (C)\int_{X^{-1}(B)} f\circ X\,\mathrm{d}\mu.$$

Proof Indeed,

$$\begin{aligned}
(C)\int_{\mathbb{R}} f\,\mathrm{d}\mu_X &= \int_0^{+\infty} \mu_X(\{t : f(t)\geq\alpha\})\,\mathrm{d}\alpha \\
&\quad + \int_{-\infty}^{0} [\mu_X(\{t : f(t)\geq\alpha\}) - \mu_X(\mathbb{R})]\,\mathrm{d}\alpha \\
&= \int_0^{+\infty} \mu(X^{-1}(\{t : f(t)\geq\alpha\}))\,\mathrm{d}\alpha \\
&\quad + \int_{-\infty}^{0} [\mu(X^{-1}(\{t : f(t)\geq\alpha\})) - \mu\left(X^{-1}(\mathbb{R})\right)]\,\mathrm{d}\alpha \\
&= \int_0^{+\infty} \mu(\{\omega : (f\circ X)(\omega)\geq\alpha\})\,\mathrm{d}\alpha \\
&\quad + \int_{-\infty}^{0} [\mu(\{\omega : (f\circ X)(\omega)\geq\alpha\}) - \mu(\Omega)]\,\mathrm{d}\alpha \\
&= (C)\int_{\Omega} f\circ X\,\mathrm{d}\mu.
\end{aligned}$$

The second assertion of the lemma is a consequence of the above discussion, by replacing μ by its restriction to the σ-algebra $\Sigma_B = \{A\cap B : A\in\Sigma\}$. ■

Given a normalized capacity space (Ω, Σ, μ) and a real interval I (bounded or unbounded), we will refer to the mappings $Z : \mathbb{N}\times I \to \mathcal{M}(\Omega, \Sigma)$ as *random schemes* on I. In what follows, the expected value and the variance of the random variables $Z(n, x) \in \mathcal{M}(\Omega, \Sigma)$ will be denoted

$$\alpha_{n,x} = \mathrm{E}_C(Z(n,x)) \quad \text{and} \quad \sigma_{n,x}^2 = \mathrm{Var}_C(Z(n,x))$$

respectively. Following Feller [63], one can associate to every random scheme $Z = Z(n, x)$ a sequence of operators $T_n : C_b(\mathbb{R}) \to \mathcal{F}_b(\mathbb{R})$ defined by

$$T_n(f)(x) = (\mathrm{C})\int_{\Omega} f\circ Z(n,x)\,\mathrm{d}\mu = (\mathrm{C})\int_{\mathbb{R}} f(t)\,\mathrm{d}\mu_{Z(n,x)}(t), \quad x\in\mathbb{R},\ n\in\mathbb{N}.$$

It is worth noticing that these operators are weakly nonlinear, monotone increasing, and invariant to translations when μ is a strongly subadditive capacity.

The nice limiting behavior of these operators is described by the following result, whose proof is inspired by the argument presented by Altomare and Campiti [12] for their Theorem 5.2.2.

Theorem 5.9.2 (The Generalization of Feller's Theorem) *Suppose that μ is a normalized and strongly subadditive capacity and $Z : \mathbb{N} \times I \to \mathcal{M}(\Omega, \Sigma)$ is a random scheme such that*

$$\lim_{n\to+\infty} \alpha_{n,x} = x \quad \textit{uniformly on } I \tag{5.29}$$

and

$$\lim_{n\to+\infty} \sigma^2_{n,x} = 0 \quad \textit{uniformly on } I. \tag{5.30}$$

Then the sequence $(T_n)_n$, of operators associated to the random scheme Z, has the limiting property

$$\lim_{n\to\infty} T_n(f) = f \quad \textit{uniformly on the compact subintervals of } I \tag{5.31}$$

for every $f \in C_b(\mathbb{R})$. Moreover, if $f \in \mathcal{U}C_b(\mathbb{R})$, then

$$\lim_{n\to\infty} T_n(f) = f \quad \textit{uniformly on } I. \tag{5.32}$$

Proof Notice first that the integrals defining $T_n(f)(x)$ are finite, for all $x \in I$ and $n \in \mathbb{N}$. Indeed, denoting $H(\alpha) = \mu(\{\omega \in \Omega : f(Z(n,x)(\omega)) \geq \alpha\})$, it is clear that $H(\alpha)$ is a finite and nonincreasing function on $\mathbb{R}$ and since f is bounded, there exist $m < 0$ and $M > 0$, such that

$$(\mathrm{C})\int_\Omega f \circ Z(n,x)\,\mathrm{d}\mu = \int_0^M H(\alpha)\,\mathrm{d}\alpha + \int_m^0 (H(\alpha)-1)\,\mathrm{d}\alpha,$$

which obviously is a finite number. Another useful remark is the fact that

$$\overline{\mu}(A) = \mu(\Omega) - \mu(\Omega \setminus A) \leq \mu(A) \quad \text{for every } A \in \Sigma,$$

due to the subadditivity of μ.

Consider next the case of uniformly continuous and bounded functions f defined on I. Then for every $\varepsilon > 0$, there exists $\delta > 0$, such that for all $x, y \in I$ with $|x-y| < \delta$, we have $|f(x) - f(y)| \leq \varepsilon/4$. According to the hypotheses (5.29) and (5.30), there exists

$N \in \mathbb{N}$ such that for every $n \geq N$ and $x \in I$ we have

$$\left|x - \alpha_{n,x}\right| \leq \delta$$

and

$$\sigma_{n,x}^2 \leq \frac{\delta^2 \varepsilon}{8\left(\|f\| + 1\right)}.$$

According to the capacitary inequality of the modulus (see Theorem 3.1.2),

$$\begin{aligned} |T_n(f)(x) - f(\alpha_{n,x})| &= \left|(\mathrm{C})\int_{\mathbb{R}} (f(t) - f(\alpha_{n,x}))\, \mathrm{d}\mu_{Z(n,x)}(t)\right| \\ &\leq (\mathrm{C})\int_{\mathbb{R}} |f(t) - f(\alpha_{n,x})|\, \mathrm{d}\mu_{Z(n,x)}(t). \end{aligned} \tag{5.33}$$

In order to evaluate the right-hand side, let us consider the decomposition

$$\mathbb{R} = \{t \in \mathbb{R} : |t - \alpha_{n,x}| < \delta\} \cup \{t \in \mathbb{R} : |t - \alpha_{n,x}| \geq \delta\}$$

denoted as $\mathbb{R} = X_1 \cup X_2$. Put

$$\Omega_{X_1} = Z^{-1}(n,x)(X_1) \text{ and } \Omega_{X_2} = Z^{-1}(n,x)(X_2).$$

Then, the following inequalities hold:

$$(\mathrm{C})\int_{X_1} 1\, \mathrm{d}\mu_{Z(n,x)}(t) = (\mathrm{C})\int_{\Omega_{X_1}} 1 \cdot d\mu \leq (\mathrm{C})\int_{\Omega} 1\, \mathrm{d}\mu = \mu(\Omega) = 1,$$

and

$$\begin{aligned} (\mathrm{C})\int_{X_2} |f(t) - f(\alpha_{n,x})|\, \mathrm{d}\mu_{Z(n,x)}(t) &\leq 2\|f\| \cdot (\mathrm{C})\int_{X_2} 1\, \mathrm{d}\mu_{Z(n,x)}(t) \\ &= 2\|f\| \cdot (\mathrm{C})\int_{\Omega_{X_2}} 1\, \mathrm{d}\mu = 2\|f\| \cdot \mu(\Omega_{X_2}) \\ &= 2\|f\| \cdot \mu(\{\omega \in \Omega : |Z(n,x)(\omega) - \alpha_{n,x}| \geq \delta\}). \end{aligned}$$

Since μ is subadditive, it follows that $\mu_{Z(n,x)}$ is also subadditive. Therefore, taking into account Theorem 2.6.5, for every $n \geq N$ and every x, the following chain of inequalities hold:

$$
\begin{aligned}
(\mathrm{C})\int_{\mathbb{R}} &|f(t)-f(\alpha_{n,x})|\,\mathrm{d}\mu_{Z(n,x)}(t)\\
&=\int_0^\infty \mu_{Z(n,x)}(\{t\in X_1\cup X_2 : |f(t)-f(\alpha_{n,x})|\geq \alpha\})\,\mathrm{d}\alpha\\
&\leq \int_0^\infty \mu_{Z(n,x)}(\{t\in X_1 : |f(t)-f(\alpha_{n,x})|\geq \alpha\})\,\mathrm{d}\alpha\\
&\quad+\int_0^\infty \mu_{Z(n,x)}(\{t\in X_2 : |f(t)-f(\alpha_{n,x})|\geq \alpha\})\,\mathrm{d}\alpha\\
&=(\mathrm{C})\int_{X_1}|f(t)-f(\alpha_{n,x})|\,\mathrm{d}\mu_{Z(n,x)}(t)+(\mathrm{C})\int_{X_2}|f(t)-f(\alpha_{n,x})|\,\mathrm{d}\mu_{Z(n,x)}(t)\\
&\leq \frac{\varepsilon}{4}\cdot(\mathrm{C})\int_{X_1}1\,\mathrm{d}\mu_{Z(n,x)}(t)+2\|f\|\cdot\mu(\{\omega : |Z(n,x)-\alpha_{n,x}|\geq\delta\})\\
&\leq \frac{\varepsilon}{4}+2\|f\|\cdot\sigma_{n,x}^2\cdot\delta^{-2}\leq\frac{\varepsilon}{4}+\frac{\varepsilon}{4}=\frac{\varepsilon}{2},
\end{aligned}
$$

where the last step is motivated by Chebyshev's inequality 3.1.5 and our choice of N_ε. The proof ends in this case by noticing that

$$
\begin{aligned}
|T_n(f)(x)-f(x)| &\leq \left|T_n(f)(x)-f(\alpha_{n,x})\right|+\left|f(\alpha_{n,x})-f(x)\right|\\
&\leq \frac{\varepsilon}{2}+\frac{\varepsilon}{4}<\varepsilon
\end{aligned}
$$

for all $n\geq N_\varepsilon$ and all $x\in\mathbb{R}$.

Consider now the case of functions $f\in C_b(\mathbb{R})$ and choose arbitrarily a number $\varepsilon>0$ such that $\varepsilon<6(\|f\|+1)$. Then for every compact subinterval $K\subset I$, one can choose a continuous function $g:\mathbb{R}\to\mathbb{R}$ with compact support such that

$$
0\leq g\leq 1 \text{ and } g|_K\geq 1-\frac{\varepsilon}{6(\|f\|+1)}.
$$

Clearly, fg and $1-g$ belong to $\mathcal{UC}_b(\mathbb{R})$, so by the above reasoning there exists a natural number N' such that for every $n\geq N'$ and every $x\in\mathbb{R}$,

$$
|T_n(1-g)(x)-(1-g(x))|\leq\frac{\varepsilon}{6(\|f\|+1)}
$$

and

$$
|T_n(fg)(x)-f(x)g(x)|\leq\frac{\varepsilon}{3}.
$$

Then, for every $x\in K$,

$$|T_n(1-g)(x)| \leq \frac{\varepsilon}{6(\|f\|+1)} + |1-g(x)| \leq \frac{\varepsilon}{3(\|f\|+1)}.$$

As a consequence, for every $x \in K$ and $n \geq N'$,

$$\begin{aligned}
|T_n(f)(x) - f(x)| &\leq |T_n(f)(x) - T_n(fg)(x)| + |T_n(fg)(x) - f(x)g(x)| \\
&\quad + |f(x)\,(1-g(x))| \\
&\leq (\mathrm{C})\int_{\mathbb{R}} |f|\,(1-g)\;\mathrm{d}\mu_{Z(n,x)} + \frac{\varepsilon}{3} + \frac{\|f\|\,\varepsilon}{6(\|f\|+1)} \\
&\leq \|f\|\left((\mathrm{C})\int_{\mathbb{R}} (1-g)\;\mathrm{d}\mu_{Z(n,x)}\right) + \frac{\varepsilon}{2} \\
&\leq \|f\|\,T_n\,(1-g)\,(x) + \frac{\varepsilon}{2} < \varepsilon
\end{aligned}$$

where the second inequality is motivated by Theorem 3.1.2 (the capacitary inequality of the modulus). ■

Remark 5.9.3 An inspection of the proof of Theorem 5.9.2 easily shows that if the conditions (5.29) and (5.30) hold uniformly only on the compact subintervals of I, then the conclusion holds uniformly on the compact subintervals of I.

Remark 5.9.4 It is worth noting that if $I \subsetneq \mathbb{R}$ is an interval and

$$\mu(\{\omega \in \Omega : Z(n,x)(\omega) \notin I\} = 0, \tag{5.34}$$

then to every sequence of operators $T_n : C_b(\mathbb{R}) \to \mathcal{F}_b(\mathbb{R})$ as in the statement of Theorem 5.9.2, one can associate operators $T_n^* : C_b(I) \to \mathcal{F}_b(I)$ such that

$$T_n^*(f) = T_n(f^*)_{|I},$$

where $f^* : \mathbb{R} \to \mathbb{R}$ is a continuous and bounded extension of f.

Indeed, due to the subadditivity of the capacity μ, the condition 5.34 implies that $\mu(\omega \in \Omega : Z(n,x)(\omega) \in I) = 1$ and thus $\mu_{Z(n,x)}$ is a normalized and subadditive capacity. On the other hand, for every subadditive capacity $\mu : \Sigma \to \mathbb{R}_+$ and every $A, B \in \Sigma$ with $\mu(B) = 0$, we have

$$\mu(A) \leq \mu(A \cup B) \leq \mu(A) + \mu(B) = \mu(A).$$

Therefore, every μ-integrable and bounded function $F : A \cup B \to \mathbb{R}$ verifies the following chain of equalities,

$$
\begin{aligned}
(\mathrm{C})\int_{A\cup B} F\,\mathrm{d}\mu &= \int_0^\infty \mu(\{t\in A\cup B : F(t)\geq\alpha\})\,\mathrm{d}\alpha \\
&\quad+\int_{-\infty}^0 [\mu(\{t\in A\cup B : F(t)\geq\alpha\})-\mu(A\cup B)]\,\mathrm{d}\alpha \\
&= \int_0^\infty \mu(\{t\in A : F(t)\geq\alpha\})\,\mathrm{d}\alpha \\
&\quad+\int_{-\infty}^0 [\mu(\{t\in A : F(t)\geq\alpha\})-\mu(A)]\,\mathrm{d}\alpha \\
&= (\mathrm{C})\int_A F\,\mathrm{d}\mu.
\end{aligned}
$$

By applying this for $A = I,\ B = \mathbb{R}\setminus I,\ F = f^*$ and $\mu = \mu_{Z(n,x)}$, we conclude that

$$(\mathrm{C})\int_{\mathbb{R}} f^*\,\mathrm{d}\mu_{Z(n,x)} = (\mathrm{C})\int_I f\,\mathrm{d}\mu_{Z(n,x)}.$$

In order to show how the Bernstein approximation theorem can be recovered from Feller's theorem, we shall need the following auxiliary result.

Lemma 5.9.5 *Let* $(\mathbb{R},\Sigma,\mu)$ *be a normalized capacity space such that there exist a set* $U=\{u_0,u_1,\ldots,u_n\}\subset\mathbb{R}$ *and nonnegative numbers* $p_0,p_1,\ldots,p_n$ *that sum to* 1 *such that*

$$\mu\left(\{u_k\}\right) = p_k \quad \textit{for } k\in\{1,\ldots,n\}$$

and

$$\mu(A) = \sum_{a\in A\cap U}\mu(\{a\}) \quad \textit{for every } A\in\Sigma.$$

Then the integral of every function $g\in C_b(\mathbb{R})$ *verifies the formula*

$$(\mathrm{C})\int_{\mathbb{R}} g\,\mathrm{d}\mu = \sum_{k=0}^n g(u_k)p_k.$$

Proof Suppose first that g is strictly positive. Then

$$(\mathrm{C})\int_{\mathbb{R}} g(t)\,\mathrm{d}\mu(t) = \int_0^\infty \mu\left(\{t\in\mathbb{R} : g(t)\geq\alpha\}\right)\,\mathrm{d}\alpha.$$

By rearranging $M=\{g(u_k) : 0\leq k\leq n\}$ as $\{m_0\leq m_1\leq\cdots\leq m_n\}$, we have

$$
\begin{aligned}
(\mathrm{C})\int_{\mathbb{R}} g(t)\,\mathrm{d}\mu(t) &= \int_{[0,m_n]} \mu\left(\{t\in\mathbb{R}:\ g(t)\geq\alpha\}\right)\,\mathrm{d}\alpha \\
&= \int_{[0,m_0]} \mu\left(\{t\in\mathbb{R}:\ g(t)\geq\alpha\}\right)\,\mathrm{d}\alpha \\
&\quad + \sum_{k=0}^{n-1}\int_{(m_k,m_{k+1}]} \mu\left(\{t\in\mathbb{R}:\ g(t)\geq\alpha\}\right)\,\mathrm{d}\alpha \\
&= m_0\mu(U) + \sum_{k=0}^{n-1}(m_{k+1}-m_k)\mu(\{u_{k+1},\ldots,u_n\}) \\
&= \sum_{k=0}^{n} m_k p_k = \sum_{k=0}^{n} g(u_k)p_k.
\end{aligned}
$$

In the general case, since g is a bounded function, there exists $m\in\mathbb{R}$, such that $g(t)>m$, for all $t\in\mathbb{R}$. Then

$$
\begin{aligned}
(\mathrm{C})\int_{\mathbb{R}} g(t)\,\mathrm{d}\mu(t) &= (\mathrm{C})\int_{\mathbb{R}}(g(t)-m)\,\mathrm{d}\mu(t) + m\cdot(\mathrm{C})\int_{\mathbb{R}}\mathrm{d}\mu(t) \\
&= \sum_{k=0}^{n}(g(u_k)-m)p_k + m\sum_{k=0}^{n}p_k = \sum_{k=0}^{n} g(u_k)p_k
\end{aligned}
$$

and the proof is complete. ■

The following example shows how Bernstein's approximation theorem can be derived from Theorem 5.9.2.

Example 5.9.6 Let (Ω,Σ,μ) be a normalized capacity space and consider the random scheme $Z:\mathbb{N}\times[0,1]\to\mathcal{M}(\Omega,\Sigma)$ such that the range of $Z(n,x)$ is included in $M_n=\left\{\frac{0}{n},\frac{1}{n},\ldots,\frac{n}{n}\right\}$ and $\mu_{Z(n,x)}$ is given by

$$
\mu_{Z(n,x)}\left(\left\{\frac{k}{n}\right\}\right) = p_{n,k}(x) = \binom{n}{k}x^k(1-x)^{n-k}
$$

for $k\in 0,1,\ldots,n$ and

$$
\mu_{Z(n,x)}(A) = \sum_{a\in A\cap M_n}\mu_{Z(n,x)}(a)\quad\text{for } A\in\mathcal{B}(\mathbb{R}),
$$

whenever $n\in\mathbb{N}$ and $x\in[0,1]$. According to Remark 5.9.4 and Lemma 5.9.5, applied for

$$\Sigma = \mathcal{B}(\mathbb{R}),\ I = [0, 1],\ \mu = \mu_{Z(n,x)},\ p_k = p_{n,k}(x),\ u_k = \frac{k}{n} \text{ and } g = f^\star,$$

the operators $T_n^\star$ constructed by this scheme coincide with the classical Bernstein operators.

To prove this assertion, notice that for all $f \in C_b([0, 1])$, $n \in \mathbb{N}$ and $x \in [0, 1]$ we have

$$\begin{aligned} T_n^\star(f)(x) &= (C)\int_I f(t)\, \mathrm{d}\mu_{Z(n,x)}(t) = \int_{\mathbb{R}} f^\star(t)\, \mathrm{d}\mu_{Z(n,x)}(t) \\ &= \sum_{k=0}^{n} f\left(\frac{k}{n}\right)\binom{n}{k} x^k(1-x)^{n-k} = B_n(f)(x) \end{aligned}$$

and also

$$\begin{aligned} \alpha_{n,x} &= (C)\int_{M_n} t\, \mathrm{d}\mu_{Z(n,x)}(t) = \sum_{k=0}^{n} \frac{k}{n} p_{n,k}(x) = x, \\ \sigma_{n,x}^2 &= (C)\int_{M_n} (t-x)^2\, \mathrm{d}\mu_{Z(n,x)}(t) \\ &= \sum_{k=0}^{n} \left(\frac{k}{n} - x\right)^2 p_{n,k}(x) = \frac{x(1-x)}{n}, \end{aligned}$$

according to the combinatorial formulas mentioned in the introduction to this chapter. Therefore, in the particular case of Example 5.9.6, the conclusion of Theorem 5.9.2 coincides with Bernstein's approximation theorem.

Many other important applications of Theorem 5.9.2 in the case of probability spaces can be found in the aforementioned book of Altomare and Campiti.

In the general case of normalized capacity spaces, the computation/estimation of $\mathrm{E}(Z(n,x))$ and $\mathrm{Var}(Z(n,x))$ can be rather complicated. See Gal [67, Example 4.1]. He proposed the following quantitative result as an alternative to Theorem 5.9.2.

Theorem 5.9.7 *Let Σ be a σ-algebra of subsets of $\mathbb{R}$, $\mu : \Sigma \to \mathbb{R}_+$ a subadditive capacity and I an interval. Consider also a family of functions $\lambda_{n,x} : \mathbb{R} \to \mathbb{R}_+$ ($n \in \mathbb{N}$, $x \in I$), Choquet integrable with respect to μ, that verify the condition*

$$(\mathrm{C})\int_I \lambda_{n,x}(t)\, \mathrm{d}\mu(t) = 1 \quad \textit{for all } n \textit{ and } x.$$

Define by $\widetilde{\mathcal{UC}}(\mathbb{R})$ the class of all uniformly continuous functions $f : \mathbb{R} \to \mathbb{R}_+$ such that all products $f \cdot \lambda_{n,x}$ are Σ-measurable and

$$T_n(f)(x) = (\mathrm{C})\int_I f(t)\lambda_{n,x}(t)\,\mathrm{d}\mu(t) < \infty \quad \textit{for all } n \textit{ and } x.$$

Then, denoting by φ_x *the function defined by the formula* $\varphi_x(t) = |t - x|$, *the following estimate holds,*

$$|T_n(f)(x) - f(x)| \leq \left[1 + \frac{T_n(\varphi_x)(x)}{\delta}\right] \cdot \omega(f, \delta),$$

whenever $f \in \widetilde{\mathcal{UC}}(\mathbb{R})$, $x \in \mathbb{R}$, $n \in \mathbb{N}$ *and* $\delta > 0$.

Proof Clearly, T_n is a monotone increasing and sublinear operator on the space $\widetilde{\mathcal{UC}}(\mathbb{R})$. Notice also that if $f, g \in \widetilde{\mathcal{UC}}(\mathbb{R})$ then $f + g \in \widetilde{\mathcal{UC}}(\mathbb{R})$ and $\alpha f \in \widetilde{\mathcal{UC}}(\mathbb{R})$, for all $\alpha \geq 0$. Moreover,

$$|T_n(f)(x) - T_n(g)(x)| \leq T_n(|f - g|)(x) \quad \text{for all } n \text{ and } x.$$

Taking into account that $T_n(1) = 1$, we infer that

$$|T_n(f)(x) - f(x)| \leq T_n(|f - f(x)|)(x)$$

On the other hand,

$$|f(t) - f(x)| \leq \omega(f, |t - x|) \leq \left[1 + \frac{1}{\delta}|t - x|\right]\omega(f, \delta),$$

so

$$|T_n(f)(x) - f(x)| \leq \left[1 + \frac{1}{\delta}T_n(\varphi_x)(x)\right]\omega(f, \delta),$$

and the proof is complete. ∎

Notice that the class $\widetilde{\mathcal{UC}}(\mathbb{R})$ appearing in Theorem 5.9.7 contains all uniformly continuous and bounded functions $f : \mathbb{R} \to \mathbb{R}_+$.

Theorem 5.9.7 will be illustrated by the mean of possibilistic Picard–Choquet operators, following the paper [67] by Gal. Before entering the details we need a preparation.

The Picard–Choquet operators attached to a strongly subadditive capacity μ, are defined on suitable classes of continuous and bounded functions $f : \mathbb{R} \to \mathbb{R}_+$ via the formula

$$T_n(f)(x) = \frac{1}{c(n, x)} \cdot (\mathrm{C})\int_{\mathbb{R}} f(t)e^{-n|x-t|}\,\mathrm{d}\mu(t)$$

$$= \frac{1}{c(n,x)} \cdot \int_0^{\infty} \mu(\{t : f(t)e^{-n|x-t|} \geq \alpha\})\, d\alpha,$$

where $c(n,x) = \int_0^{\infty} \mu\left(\{t \in \mathbb{R} : e^{-n|x-t|} \geq \alpha\}\right) d\alpha$. Clearly,

$$F_\alpha = \{t \in \mathbb{R} : e^{-n|x-t|} \geq \alpha\} = \emptyset \quad \text{for } \alpha > 1$$

and for $0 < \alpha \leq 1$ we have

$$\begin{aligned} F_\alpha &= \{t \in \mathbb{R} : t \geq x,\ e^{-n(t-x)} \geq \alpha\} \bigcup \left\{t \in \mathbb{R} : t < x,\ e^{-n(x-t)} \geq \alpha\right\} \\ &= \left\{t \in \mathbb{R} : x \leq t \leq \frac{nx - \ln\alpha}{n}\right\} \bigcup \left\{t \subset \mathbb{R} : \frac{\ln(\alpha) + nx}{n} \leq t \leq x\right\} \\ &= \left[\frac{nx + \ln\alpha}{n}, \frac{nx - \ln\alpha}{n}\right]. \end{aligned}$$

As a consequence,

$$c(n,x) = \int_0^1 \mu\left(\left[x + \frac{\ln(\alpha)}{n}, x - \frac{\ln(\alpha)}{n}\right]\right) d\alpha.$$

Example 5.9.8 (The Case of possibilistic Picard–Choquet Operators) It has already been noted that the possibility measures are strongly subadditive capacities. In the case of the possibility measures defined on $\mathcal{P}(\mathbb{R})$ by the formula

$$\mu_{n,x}(A) = \begin{cases} \sup\{e^{-n|t-x|} : t \in A\} & \text{if } A \neq \emptyset \\ 0 & \text{if } A = \emptyset, \end{cases}$$

the Picard–Choquet operators become

$$\begin{aligned} T_n(f)(x) &= \frac{1}{c(n,x)} \cdot (\text{C}) \int_{\mathbb{R}} f(t)e^{-n|t-x|}\, d\mu_{n,x}(t) \\ &= \frac{1}{c(n,x)} \cdot \int_0^{\infty} \sup\{e^{-n|t-x|} : t \in \mathbb{R},\ f(t)e^{-n|t-x|} \geq \alpha\}\, d\alpha \end{aligned}$$

and their convergence to f depends on the upper estimates of the quantity

$$T_n(\varphi_x)(x) = \frac{1}{c(n,x)} \cdot \int_0^{\infty} \sup\{e^{-n|t-x|} : t \in \mathbb{R},\ |t-x|e^{-n|t-x|} \geq \alpha\}\, d\alpha.$$

By the above formula of $c(n,x)$, we obtain

$$\begin{aligned} c(n,x) &= \int_0^\infty \sup\{e^{-n|t-x|} : t \in \mathbb{R}, e^{-n|t-x|} \ge \alpha\}\, d\alpha \\ &= \int_0^1 \sup\{e^{-n|t-x|} : t \in [x + \ln(\alpha)/n, x - \ln(\alpha)/n]\}\, d\alpha. \end{aligned}$$

Since $t \in [x + (\ln\alpha)/n, x - (\ln\alpha)/n]$ is equivalent with $0 \le |t - x| \le -(\ln\alpha)/n$, it follows that

$$c(n,x) = \int_0^1 \sup\{e^{-n|t-x|} : t \in [x + \ln(\alpha)/n, x - \ln(\alpha)/n]\}\, d\alpha = \int_0^1 d\alpha = 1.$$

By Theorem 5.9.7, for all $f \in \widetilde{\mathcal{U}C}(\mathbb{R})$, $n \in \mathbb{N}$, $x \in \mathbb{R}$, and $\delta > 0$, we get

$$|T_n(f)(x) - f(x)| \le \left[1 + \frac{1}{\delta} T_n(\varphi_x)(x)\right] \omega(f;\delta), \tag{5.35}$$

where $\varphi_x(t) = |t - x|$. Therefore, the convergence of $T_n(f)$ to f relies on the convergence to zero, as $n \to \infty$, of the quantity

$$\begin{aligned} T_n(\varphi_x)(x) &= \frac{1}{c(n,x)} \cdot \int_0^\infty \sup\{e^{-n|t-x|} : t \in \mathbb{R},\ |t - x| e^{-n|t-x|} \ge \alpha\}\, d\alpha \\ &= \int_0^\infty \sup\{e^{-ny} : y \ge 0,\ ye^{-ny} \ge \alpha\}\, d\alpha. \end{aligned}$$

Now, denoting $F(v) = ve^{-nv}$, $v \ge 0$, we have $F'(v) = (1 - nv)e^{-nv}$, which immediately implies that $v = \frac{1}{n}$ is a point of maximum for F on $[0, +\infty)$ and $F(1/n) = (ne)^{-1}$ is the maximum value for F. This implies that for $\alpha > 1/(ne)$, we have $\{t \in \mathbb{R} : |t - x|e^{-n|t-x|} \ge \alpha\} = \emptyset$ and therefore,

$$T_n(\varphi_x)(x) = \int_0^{1/(ne)} \sup\{e^{-ny} : y \ge 0,\ ye^{-ny} \ge \alpha\}\, d\alpha \le \int_0^{1/(ne)} d\alpha = \frac{1}{ne}.$$

Then, choosing $\delta = (ne)^{-1}$ in Theorem 5.9.7, we get the estimate

$$|T_n(f)(x) - f(x)| \le 2\omega\left(f; \frac{1}{ne}\right),$$

which assures the convergence in the case of all uniformly continuous and bounded functions f.

5.10 Comments

Korovkin's theorem is sometimes mentioned in the literature as the Popoviciu–Bohman–Korovkin theorem or the Bohman–Korovkin theorem. Since the contribution of Popoviciu [155] was published in Romanian, it remained unnoticed for a long time. Bohman [26] considered only the case of positive linear operators $T_n : C([a, b]) \to C([a, b])$ of the form

$$T_n(f)(x) = \sum_{i=0}^{n} u_{n,i}(x) f(x_{n,i}),$$

but his proof also covers Korovkin's theorem in full generality.

A nice account on the general theory of approximation of continuous functions on various spaces is offered by the book of Bucur and Păltineanu [29].

All material in Sect. 5.1 (including Theorem 5.1.1) is borrowed from Gal and Niculescu [75], while the theory of separating functions presented in Sect. 5.2 and the generalization of a theorem of Altomare presented in Sect. 5.3 follow closely the paper [78] by Gal and Niculescu.

The variants of Theorem 5.1.1 for almost everywhere convergence, convergence in measure and convergence in p-mean (as presented in Sects. 5.4–5.6) are based on the paper [79].

Notice the existence of a large literature concerning Korovkin theory, produced by replacing the requirement of pointwise or uniform convergence by statistical convergence. Starting with the seminal paper of Gadjiev and Orhan [66], who opened this direction of research, hundreds of new contributions have appeared in a large variety of contexts. Some generalizations to the case of monotone increasing and sublinear operators can be found in the papers of Gal and Iancu [73] and Iancu [100]. Some historical information about the origin of statistical convergence can be found in a series of papers by Niculescu and Popovici [138–140].

The operator variants of Korovkin's theorem presented in Sect. 5.7 follow the work by Gal and Niculescu [81], while the accompanying quantitative approximation results from Sect. 5.8 were published by the same two authors in [80].

Several important quantitative approximation results in the case of positive and linear operators acting on the space $L^p([0, 1])$ with $1 \leq p \leq +\infty$ have been published in the 80s by Berens and DeVore [22] and Swetits and Wood [175]. A first attempt to extend their contribution to a nonlinear context was made by Gal and Niculescu [82], but some details remain to be clarified.

The extension of Feller's scheme of approximation in Sect. 5.9 is due to Gal [67].

References

1. Abramovich, Y.A., Aliprantis, C.D.: Positive operators. In: Handbook of the Geometry of Banach Spaces, I, pp. 85–122. North-Holland, Amsterdam (2001)
2. Adams, D.R.: Choquet integrals in potential theory. Publ. Mat. **42**, 3–66 (1998)
3. Adams, D.R., Hedberg, L.I.: Function spaces and potential theory. Grundlehren der Mathematischen Wissenschaften, vol. 314. Springer Science & Business Media, Berlin (2012)
4. Agahi, H.: A refined Hölder's inequality for Choquet integral by Cauchy-Schwarz's inequality. Inf. Sci. **512**, 929–934 (2020)
5. Aliprantis, C.D., Burkinshaw, O.: Positive operators. In: Pure and Applied Mathematics, vol. 119. Springer Science & Business Media, Berlin (2006)
6. Aliprantis, C.D., Tourky R.: Cones and duality. In: Graduate Studies in Mathematics, vol. 84. American Mathematical Society, Providence (2007)
7. Altomare, F.: Korovkin-type theorems and positive operators. Surv. Approx. Theory **6**, 92–164 (2010)
8. Altomare, F.: On positive linear functionals and operators associated with generalized means. J. Math. Anal. Appl. **502**, paper no. 125278 (2021)
9. Altomare, F.: On the convergence of sequences of positive linear operators and functionals on bounded function spaces. Proc. Amer. Math. Soc. **149**, 3837–3848 (2021)
10. Altomare F.: Korovkin-type theorems and local approximation problems. Expo. Math. **40**(4), 1229–1243 (2022)
11. Altomare F.: On some operator functional equations related to evaluation functionals and weighted composition operators on function spaces. Aequat. Math. **99**, 1371–1384 (2025)
12. Altomare, F., Campiti, M.: Korovkin-type approximation theory and its applications. In: de Gruyter Studies in Mathematics, vol. 17. de Gruyter, Berlin (1994, reprinted 2011)
13. Amann, H.: Multiple positive fixed points of asymptotically linear maps. J. Funct. Anal. **17**(2), 174–213 (1974)
14. Amann, H.: Fixed point equations and nonlinear eigenvalue problems in ordered Banach spaces. SIAM Rev. **18**, 620–709 (1976)
15. Andréief, K.A.: Notes sur une relation entre les intégrales définies des produits des fonctions. Mém. Soc. Sci. Bordeaux (3) **2**, 1–14 (1883)
16. Ash, R.B.: Measure, Integration and Functional Analysis. Academic Press, New York (1972)
17. Aumann, R.J., Shapley L.S.: Values of Non-atomic Games. Princeton University Press, Princeton (2015)
18. Bartle, R.G., Dunford, N., Schwartz, J.: Weak compactness and vector measures. Canad. J. Math. **7**, 289-305 (1955)

S. G. Gal, C. P. Niculescu, *Choquet Integral and Monotone Sublinear Operators*,
Frontiers in Mathematics, https://doi.org/10.1007/978-3-031-71701-7

19. Bassanezi, R.C., Greco, G.H.: Sull'additività dell'integrale. Rendiconti Seminario Matematico Università di Padova **72**, 249–275 (1984)
20. Bede, B., Coroianu L., Gal, S.G.: Approximation by Max-Product Type Operators. Springer, Switzerland (2016)
21. Berens H., DeVore, R.A.: Quantitative Korovkin theorems for L^p-spaces. In: Lorentz, G.G. (eds.) Approximation Theory II, pp. 289–298. Academic Press, New York (1976)
22. Berens, H., DeVore, R.A.: Quantitative Korovkin theorems for positive linear operators in L^p space. Trans. Amer. Math. Soc. **245**, 349–361 (1978)
23. Bernstein, S.N.: Démonstration du théorème de Weierstrass fondeé sur le calcul des probabilités. Commun. Soc. Math. Kharkov Sér. 2 **13**, 1–2 (1912/1913)
24. Bhatia, R.: Notes on functional analysis. In: Texts and Readings in Mathematics, vol. 50. Hindustan Book Agency, New Delhi (2009)
25. Bogachev, V.I.: Measure Theory, vol. I. Springer Science & Business Media, Berlin (2007)
26. Bohman, H.: On approximation of continuous and of analytic functions. Ark. Mat., **2**, 43–56 (1952–1954)
27. Brodskii, M.S.: On a problem of I.M. Gelfand. Uspekhi Mat. Nak. **12**, 129–132 (1957)
28. Brodskii, M.S.: Triangular and Jordan representation of linear operators (translated from Russian by J. M. Danskin), Translations of Mathematical Monographs, vol. 32. American Mathematical Society, Providence (1971)
29. Bucur I., Păltineanu, G.: Topics in Uniform Approximation of Continuous Functions. Birkhäuser, Basel (2020)
30. Buskes, G.: The Hahn-Banach theorem surveyed. Dissertationes Math. (Rozprawy Mat.) **327**, 1–49 (1993)
31. Carleson, L.: Selected Problems on Exceptional Sets. D. Van Nostrand Company, Princeton (1967)
32. Carothers, N.L.: Real Analysis. Cambridge University Press, Cambridge (2000)
33. Cerdà, J., Coll, H., Martín, J.: Entropy function spaces and interpolation. J. Math. Anal. Appl. **304**, 269–295 (2005)
34. Cerdà, J., Martín, J., Silvestre, P.: Capacitary function spaces. Collect. Math. **62**, 95–118 (2011)
35. Cerreia-Vioglio, S., Maccheroni, F., Marinacci, M., Montrucchio, L.: Signed integral representations of comonotonic additive functionals, J. Math. Anal. Appl. **385**(2), 895–912 (2012)
36. Cerreia-Vioglio, S., Maccheroni, F., Marinacci, M., Montrucchio, L.; Choquet integration on Riesz spaces and dual comonotonicity. Trans. Amer. Math. Soc. **367**, 8521–8542 (2015)
37. Chateauneuf, A.: On the use of capacities in modeling uncertainty aversion and risk aversion. J. Math. Econom. **20**, 343–369 (1991)
38. Chateauneuf, A., Lefort, J.-P.: Some Fubini theorems on product σ-algebras for non-additive measures. Int. J. Approx. Reason **48**, 686–696 (2008)
39. Chebyshev, P.L.: Des valeurs moyennes. J. Math. Pures Appl. 2 **12**, 177–184 (1867)
40. Chebyshev, P.L.: On approximate expressions of some integrals in terms of others, taken within the same limits (Russian). Proc. Math. Soc. Karkov **2**, 93–98 (1882)
41. Choquet, G.: Theory of capacities. Ann. Inst. Fourier (Grenoble) **5**, 131–292 (1953–1954)
42. Choquet, G.: La naissance de la théorie des capacités: réflexion sur une expérience personnelle. Comptes rendus de l'Académie des sciences, Série générale, La Vie des sciences **3**, 385–397 (1986)
43. Choudary, A.D.R., Niculescu, C.P.: Real Analysis on Intervals. Springer India, New Delhi (2014)
44. Conway, J.B.: A Course in Operator Theory. American Mathematical Society, Providence (2000)
45. Crandall M.G., Tartar L.: Some relations between nonexpansive and order preserving maps. Proc. Amer. Math. Soc. **78**, 385–390 (1980)

46. Davies, E.B.: The structure and ideal theory of the pre-dual of a Banach lattice. Trans. Amer. Math. Soc. **131**, 544–555 (1968)
47. De Campos, L.M., Bolaños, M.J.: Characterization and comparison of Sugeno and Choquet integrals. Fuzzy Sets Syst. **52**(1), 61–67 (1992)
48. Dekking, F.M., Kraaikamp, C., Lopuhaä, H.P., Meester, L.E.: A Modern Introduction to Probability and Statistics: Understanding Why and How. Springer Science & Business Media, Berlin (2006)
49. Dellacherie, Cl.: Quelques commentaires sur les prolongements de capacités. Séminaire de Probabilités V, Strasbourg, Lecture Notes in Mathematics, vol. 191, pp. 77–81. Springer-Verlag, Berlin and New York (1971)
50. Dempster, A.: Upper and lower probability induced by a multi-valued mapping. Ann. Math. Stat. **38**, 325–339 (1967)
51. Denneberg, D.: Non-Additive Measure and Integral. Kluwer Academic Publishers, Dordrecht (1994)
52. Diestel, J., Uhl, J.J.: Vector Measures. Mathematical Surveys No. 15, American Mathematical Society, Providence (1977)
53. Dixmier, J.: Les opérateurs permutables à l'opérateur integral. Portugal Math. **8**, 73–84 (1949)
54. Donner K.: Korovkin's theorem in L^p-spaces. J. Funct. Anal. **42**, 12–28 (1981)
55. Donner, K.: Extension of Positive Operators and Korovkin Theorems. Springer, Berlin (2006)
56. Donoghue, W.F.: The latice of invariant subspaces of quasi-nilpotent completely continuous transformation. Pacific J. Math. **7**, 1031–1035 (1957)
57. Dubois, D., Prade, H.: Théorie des Possibilités. Masson, Paris (1985)
58. Epstein, L.G., Wang, T.: "Beliefs about Beliefs" without Probabilities. Econometrica **64**(6), 1343–1373 (1996)
59. Evans, L.C., Gariepy, R.F.: Measure Theory and Fine Properties of Functions. Revised edition. Studies in Advanced Mathematics, CRC Press, Boca Raton (2015)
60. Falconer, K.: Fractal Geometry: Mathematical Foundations and Applications, 2nd edn. John Wiley & Sons, New York (2013)
61. Feinstein, J.F.: Convergence from below suffices. Irish Math. Soc. Bull. **59**, 65–70 (2007); see also arXiv preprint arXiv:1412.7702 (2014)
62. Feldman, M.M.: Sublinear operators defined on a cone. Siberian Math. J. **16**, 1005–1015 (1975)
63. Feller, W.: An Introduction to Probability Theory and Its Applications, vol. II. John Wiley & Sons, New York (1966)
64. Ferguson, S.J., Wu, T.: A One-Sentence Inverse Image Proof of Lusin's Theorem. See arXiv preprint arXiv:1810.13240 (2018)
65. Fichtenholz, G., Kantorovich, L.V. : Sur les opérations linéaires dans l'espace des fonctions bornées. Studia Mathematica **5**, 69–98 (1934)
66. Gadjev, A.D., Orhan, C.: Some approximation theorems via statistical convergence. Rocky Mount. J. Math. **32**(1), 129–138 (2002)
67. Gal, S.G.: Approximation by Choquet integral operators. Ann. Mat. Pura Appl. (4) **195**(3), 881–896 (2016)
68. Gal, S.G.: Uniform and pointwise quantitative approximation by Kantorovich-Choquet type integral operators with respect to monotone and submodular set functions, Mediterr. J. Math. **14**(5) (2017). Paper no. 205 (2018)
69. Gal, S.G.: Volterra-Choquet integral equations. J. Integr. Equ. Appl. **31**, no. 4, 495–504 (2019)
70. Gal, S.G.: Approximation by polynomial possibilistic integral operators. Ann. Acad. Rom. Sci. ser. Math. Appl. **12**, 132–141 (2020)
71. Gal, S.G.: Volterra-Choquet nonlinear operator, Topol. Methods Nonlin. Anal. **57**, no. 1, 89–106 (2021)

72. Gal, S.G.; Iterates of monotone and sublinear operators on spaces of continuous functions. Results Math. **78**, article no. 152 (2023)
73. Gal, S.G., Iancu, I.T.: Korovkin-type theorems for statistically convergent sequences of monotone and sublinear operators. Bull. Malays. Math. Sci. Soc. **46**, Article no. 79 (2023)
74. Gal, S.G., Niculescu, C.P.: Kantorovich's mass transport problem for capacities. Proc. Rom. Acad. Ser. A **20**(4), 337–345 (2019)
75. Gal, S.G., Niculescu, C.P.: A nonlinear extension of Korovkin's theorem. Mediterr. J. Math. **17**, Article no. 145 (2020)
76. Gal, S.G., Niculescu, C.P.: Choquet operators associated to vector capacities. J. Math. Anal. Appl. **500**, Article no. 125153 (2021)
77. Gal, S.G., Niculescu, C.P.: A note on the Choquet type operators. Aequationes Math. **95**, 433–447 (2021)
78. Gal, S.G., Niculescu, C.P.: Nonlinear versions of Korovkin's abstract theorems. Rev. Real Acad. Cienc. Exactas Fis. Nat. Ser. A-Mat. **116**, Article number 68 (2022)
79. Gal, S.G., Niculescu, C.P.: Korovkin type theorems for weakly nonlinear and monotone operators. Mediterr. J. Math. **20**, Article no. 56 (2023)
80. Gal, S.G., Niculescu, C.P., Degree of convergence in operator versions of nonlinear Korovkin's theorems. Boll. Unione Mat. Ital. **17**, 577–587 (2023)
81. Gal, S.G., Niculescu, C.P.: Nonlinear operator extensions of Korovkin's theorems. Positivity **28**, Article no. 14 (2024)
82. Gal, S.G., Niculescu, C.P.: Quantitative Korovkin theorems for sublinear, monotone and strongly translatable operators in $L_p([0, 1])$, $1 \leq p \leq \infty$. Ann. Pol. Math. **132**(2), 137–151 (2024)
83. Gallardo, E., Montes, A.: Volterra operator is not supercyclic. Integr. Equ. Oper. Theory **50**, no. 2, 211–216 (2004)
84. Ghirardato, P.: On independence for non-additive measures, with a Fubini theorem. J. Econ. Theory **73**, 261–291 (1997)
85. Gilboa, I.: Theory of Decision under Uncertainty. Cambridge University Press, Cambridge (2009)
86. Gilboa, I., Schmeidler, D.: Additive representations of non-additive measures and the Choquet integral. Ann. Oper. Res. **52**, 43–65 (1994)
87. Girotto, B., Holzer, S.: Chebyshev type inequality for Choquet integral and comonotonicity. Internat. J. Approx. Reason. **52**, 1118–1123 (2011)
88. Glück, J., Groh, U.: A note on the positivity of inverse operators acting on C^*–algebras. Linear Algebra Appl. **708**, 337–354 (2025)
89. Gohberg, I.C., Krein, M.G.: Theory and Applications of Volterra Operators in Hilbert Spaces (translated from Russian by A. Feinstein). Translations of Mathematical Monographs, vol. 24. American Mathematical Society, Providence (1970)
90. Grabisch, M.: Set Functions, Games and Capacities in Decision Making, Theory and Decision Library C (Game Theory, Social Choice, Decision Theory, and Optimization), vol. 46. Springer, Berlin (2016)
91. Graf, S.: A Radon-Nikodym theorem for capacities. J. Reine Angew. Math. **1980**, 192–214 (1980)
92. Grafakos, L.: Classical Fourier Analysis, Graduate Texts in Mathematics, vol. 249, 3rd edn. Springer Science + Business Media, New York (2014)
93. Greco, G.H.: Integrale monotono. Rend. Sem. Mat. Padova **57**, 149–166 (1977)
94. Grigoryan, A.: Measure theory and probability. Lecture Notes, University of Bielefeld, October 2007–February 2008. https://luiarthur.github.io/padawan/assets/ams241/measure/grigoryan.pdf

95. Guessab, A., Schmeisser, G.: Two Korovkin-type theorems in multivariate approximation, Banach J. Math. Anal. **2**(2), 121–128 (2008)
96. Halmos, P.R.: A Hilbert Space Problem Book. Springer Verlag, New York (1967)
97. Harjulehto, P., Hurri-Syrjänen, R.: On Hausdorff content maximal operator and Riesz potential for non-measurable functions. Preprint. arXiv:2405.12113 (2024)
98. Hewitt, E., Stromberg, K.: Real and Abstract Analysis: A Modern Treatment of the Theory of Functions of a Real Variable. Springer Science & Business Media, Berlin (2012)
99. Hildebrandt, T.H.: On bounded functional operations. Trans. Amer. Math. Soc **36**, 868–875 (1934) (1927)
100. Iancu, I.T.: Statistical Korovkin-type theorem for monotone and sublinear operators. Stud. Univ. Babeş–Bolyai, Math. **68**(2), 399–408 (2023)
101. Jessen, B.: Bemaerkinger om konvekse Funktioner og Uligheder imellem Middelvaerdier I. Mat. Tidsskrift B, 17–28 (1931)
102. Kadison, R.V.: A representation theory for commutative topological algebra. Mem. Amer. Math. Soc. **7** (1951)
103. Kalisch, G.K.: On similarity, reducing manifolds, and unitary equivalence of certain Volterra operators. Ann. Math. **66**, 481–494 (1957)
104. Kantorovich, L.V.: Sur certains développements suivant les polynômes de la forme de S. Bernstein, I, II. C.R. Acad. Sci. URSS 563–568, 595–600 (1930)
105. Kantorovich, L.V.: On semi-ordered linear spaces and their applications to the theory of linear operators. Dokl. Akad. Nauk **4**, 11–14 (1935)
106. Kiran Kumar, V.B., Vinaya, P.C.: Operator version of Korovkin theorem; degree of convergence and application to preconditioners. J. Math. Anal. Appl. **523**, art. 127087 (2023)
107. Korovkin, P.P.: On convergence of linear positive operators in the space of continuous functions (Russian). Doklady Akad. Nauk. SSSR (NS) **90**, 961–964 (1953)
108. Korovkin, P.P.: Linear Operators and Approximation Theory. Hindustan Publishing Corporation, Delhi (1960)
109. Krein, M. G.: Fundamental properties of normal conical sets in a Banach space. Dokl. Akad. Nauk USSR **28**, 13–17 (1940)
110. Kusraev, A.G., Kutateladze S.S.: Subdifferentials: Theory and Applications. Mathematics and its Applications, vol. 323, Kluwer Academic Publishers Group, Dordrecht (1995)
111. Kutateladze, S.S.: Convex Operators. Russ. Math. Surv. **34**(1), 181–214 (1979)
112. Landkof, N.S.: Foundations of Modern Potential Theory. Translated from the Russian by A. P. Doohovskoy. Die Grundlehren der mathematischen Wiss., Band, vol. 180. Springer–Verlag, New York/Heidelberg (1972)
113. Lehrer, E.: A new integral for capacities. Econ. Theory **39**, 157–176 (2009)
114. León-Saavedra, F., Piqueras-Lerena, A.: Cyclic properties of Volterra operator. Pacific J. Math. **211**(1), 157–162 (2003)
115. León-Saavedra, F., Piqueras-Lerena, A.: Super convex-cyclicity and the Volterra operator. Monatsh. Math. **177**, no. 2, 301–305 (2015)
116. Li, J., Mesiar, R., Zhang, Q.: Absolute continuity of monotone measure and convergence in measure. Commun. Comput. Inf. Sci. (CCIS) **80**, 500–504 (2010)
117. Lieb, E.H, Loss, M.: Analysis, 2nd edn. Graduate Studies in Mathematics, vol. 14, American Mathematical Society, Providence (2001)
118. Little, G., Reade, J.B.: Estimates for the norm of the n-th indefinite integral. Bull. London Math. Soc. **30**, 539–542 (1998)
119. Lomelí, H.E., García, C.L.: Variations on a theorem of Korovkin. Amer. Math. Monthly **113**(8), 744–750 (2006)

120. Lorentz, G.G.: Approximation of Functions, 2nd edn. Chelsea Publishing Company, New York (1986)
121. Lovász, L.: Submodular setfunctions on sigma-algebras, matroids and graph limits. arXiv.2302.04704v2 (2023)
122. Lovász, L.: Choquet extension of non-monotone submodular setfunctions. Acta Math. Hungar. **176**(1), 1–14 (2025)
123. Marinacci, M.: Vitali's early contribution to non-additive integration. Rivista di matematica per le scienze economiche e sociali **20**(2), 153-158 (1997)
124. Marinacci, M., Montrucchio, L.: Introduction to the mathematics of ambiguity. In: I. Gilboa (ed.), Uncertainty in Economic Theory. Routledge, New York (2004)
125. Mattila, P.: Geometry of sets and measures in Euclidean spaces: fractals and rectifiability. Cambridge Studies in Advanced Mathematics, vol. 44. Cambridge University Press, Cambridge (1999)
126. Maz'ja, V.G., Havin, V.P.: A nonlinear potential theory. Uspehi Mat. Nauk **27**(60), 67–138 (1972) (in Russian). English translation: Russ. Math. Surv. **27**, 71–148 (1972)
127. Mesiar, R.: Choquet-like integral. J. Math. Anal. Appl. **194**, 477–488 (1995)
128. Mesiar, R., Li, J., Pap, E.: The Choquet integral as Lebesgue integral and related inequalities. Kybernetika (Prague) **46**, 1098–1107 (2010)
129. Meyer-Nieberg, P.: Banach Lattices. Springer-Verlag, Berlin (1991)
130. Milgrom, P., Roberts, J.: Rationalizability, learning, and equilibrium in games with strategic complementarities. Econometrica **58**(6), 1255–1277 (1990)
131. Namioka, I.: Partially ordered linear topological spaces. Mem. Amer. Math. Soc. **24** (1957)
132. Niculescu, C.P.: Absolute continuity and weak compactness. Bull. Amer. Math. Soc. **81**, 1064–1066 (1975)
133. Niculescu, C.P.: Absolute continuity in Banach space theory. Rev. Roum. Math. Pures Appl. **24**, 413–423 (1979)
134. Niculescu, C.P.: An overview of absolute continuity and its applications. Int. Ser. Numer. Math. **157**, 201–214 (2009)
135. Niculescu, C.P.: Old and New on Strongly Subadditive/Superadditive functions. Anal. Math. Phys. **15**, article no. 100 (2025)
136. Niculescu, C.P., Olteanu, O.: From the Hahn-Banach extension theorem to the isotonicity of convex functions and the majorization theory. Rev. R. Acad. Cienc. Exactas Fís. Nat. Ser. A Math., RACSAM **114**(4), article no. 171 (2020)
137. Niculescu, C.P., Persson, L.-E.: Convex Functions and their Applications, A Contemporary Approach, 3rd edn. CMS Books in Mathematics, vol. 23. Springer-Verlag, New York (2025)
138. Niculescu, C.P., Popovici, F.: A note on the behavior of integrable functions at infinity. J. Math. Anal. Appl. **381**(2), 742–747 (2011)
139. Niculescu, C.P., Popovici, F.: The behavior at infinity of an integrable function. Expo. Math. **30**(3), 277–282 (2012)
140. Niculescu, C.P., Popovici, F.: The asymptotic behavior of integrable functions. Real Anal. Exch. **38**(1), 157–168 (2012/2013)
141. Olteanu, O.: Convexité et prolongement d'opérateurs linéaires. C. R. Acad. Sci. Paris, Série A **286**, 511–514 (1978)
142. Olteanu, O.: Théorèmes de prolongement d'opérateurs linéaires. Rev. Roumaine Math. Pures Appl. **28**(10), 953–983 (1983)
143. Ontkovičová, Z., Torra, V.: Computation of Choquet integrals: analytical approach for continuous functions. Inf. Sci. **679**, Paper no. 121105 (2024)
144. Ooi, K.H.: On the Choquet integrals associated to Bessel capacities. Czechoslovak Math. J. **72**, 433–447 (2022)

145. Ooi, K.H.: On the dual of Choquet integrals spaces associated with capacities. Studia Math. **274**(3), 249–268 (2024)
146. Ooi, K.H., Phuc, N.C.: Characterizations of predual spaces to a class of Sobolev multiplier type spaces. J. Funct. Anal. **282**(6), Paper no. 109348 (2022)
147. Ó'Searcóid, M.: Metric Spaces. Springer Science & Business Media, Berlin (2006)
148. Ouyang, Y., Li, J.: A Radon-Nikodym theorem for monotone measures. Fuzzy Sets Syst. **487**, Article no. 108995 (2024)
149. Pap, E.: Three types of generalized Choquet integral. Boll. Unione Mat. Ital. **13**, 545–553 (2020)
150. Ponce, A.C.: Elliptic PDEs, Measures and Capacities. From the Poisson Equation to Nonlinear Thomas–Fermi Problems. European Mathematical Society, Zürich (2016)
151. Ponce, A.C., Spector, D.: Some Remarks on Capacitary Integrals and Measure Theory. In: Potentials and Partial Differential Equations: The Legacy of David R. Adams, pp. 235–264. de Gruyter, Berlin (2023)
152. Popa, D.: Korovkin-type results for multivariate functions which are periodic with respect to one variable. Math. Nachr. **291**, 1563–1573 (2018)
153. Popa, D.: An operator version of Korovkin's theorem. J. Math. Anal. Appl. **515**(1), paper no. 126375 (2022)
154. Popa, D.: An operator version of the Korovkin theorem for continuous periodic functions. J. Math. Anal. Appl. **524**(2), paper no. 127085 (2023)
155. Popoviciu, T.: Asupra demonstraţiei teoremei lui Weierstrass cu ajutorul polinoamelor de interpolare, Lucrările Sesiunii Gen. Şt. Acad. Române, 2–12 iunie 1950, Editura Academiei Republicii Populare Române, 1951, pp. 1664–1667. [Translated in English by Daniela Kacsó: On the proof of Weierstrass' theorem using interpolation polynomials. East Journal on Approximations **4**(1), 107–110 (1998)]
156. Rao, K.B., Rao, M.B.: Theory of Charges: A Study of Finitely Additive Measures. Academic Press, Cambridge (1983)
157. Ridaoui, M., Grabisch, M.: Choquet integral calculus on a continuous support and its applications. Oper. Res. Decis. **26**(1), 73–93 (2016)
158. Rubinov, A.M.: Sublinear functionals defined on a cone. Siberian Math. J. **11**, 327–336 (1970)
159. Rubinov, A.M.: Sublinear operators and their applications. Russ. Math. Surv. **32**(4), 115–175 (1977)
160. Salas, N.H.: Supercyclicity and weighted shifts. Studia Math. **135**, 55–74 (1999)
161. Sarason, D.: A remark on the Volterra operator. J. Math. Anal. Appl. **12**, 244–246 (1965)
162. Schaefer, H.H.: Banach Lattices and Positive Operators. Springer-Verlag, Berlin (1974)
163. Schaefer, H.H., Wolff, M.P.: Topological Vector Spaces. Graduate Texts in Mathematics, vol. 3, 2nd edn. Springer Science+Business Media, New York (1999)
164. Schmeidler, D.: Integral representation without additivity. Proc. Amer. Math. Soc. **97**, 255–261 (1986)
165. Schmeidler, D.: Subjective probability and expected utility without additivity. Econometrica **57**, 571–587 (1989)
166. Shafer, G.: A Mathematical Theory of Evidence. Princeton University Press, Princeton (1976)
167. Shafer, G.: Allocations of probability. Ann. Probab. **7**(5), 827–839 (1979)
168. Shisha, O., Mond, B.: The degree of convergence of sequences of linear positive operators. Proc. Natl. Acad. Sci. USA **60**, 1196–1200 (1968)
169. Šipoš, J.: Integral with respect to a pre-measure. Math. Slovaca **29**(2), 141–155 (1979)
170. Šipoš, J.: Nonlinear integrals. Math. Slovaca, **29**(3), 257–270 (1979)
171. Sugeno, M.: Theory of Fuzzy Integrals and Its Applications. PhD Dissertation, Tokyo Institute of Technology, Tokyo, Japan (1974)

172. Sugeno, M.: Fuzzy measures and fuzzy integrals–a survey. In: Gupta, S., Gaines (eds.), Fuzzy Automata and Decision Processes, pp. 89–102. North Holland, Amsterdam (1977)
173. Sugeno, M.: A note on derivatives of functions with respect to fuzzy measures. Fuzzy Sets Syst. **222**, 1–17 (2013)
174. Sugeno, M.: A way to Choquet calculus. IEEE Trans. Fuzzy Syst. **23**(5), 1439–1457 (2015)
175. Swetits, J.J., Wood, B.: Quantitative estimates for L^p approximation with positive linear operators. J. Approx. Theory **38**, 81–89 (1983)
176. Torra V, Narukawa Y, Sugeno M.: Non-additive measures: theory and Applications. Studies in Fuzziness and Soft Computing, vol. 310. Springer, Berlin (2014)
177. Valadier, M.: Sous-différentiabilité de fonctions convexes à valeurs dans un espace vectoriel ordonné. Math. Scand. **30**, 65–74 (1972)
178. Vitali, G.: Sulla definizione di integrale delle funzioni di una variabile. Ann. Pura ed Appl. **4**(2), 111–121 (1925)
179. Volkov, V.I.: On the convergence of sequences of linear positive operators in the space of continuous functions of two variables (Russian). Dokl. Akad. Nauk. SSSR (N.S.) **115**, 17–19 (1957)
180. Vretblad, A.: Fourier Analysis and its Applications. Graduate Texts in Mathematics, vol. 223. Springer, New York (2000)
181. Wang, H.: Korovkin-type theorem and application. J. Approx. Theory **132**, 258–264 (2005)
182. Wang, R.-S.: Some inequalities and convergence theorems for Choquet integrals. J. Appl. Math. Comput. **35**, 305–321 (2011)
183. Wang, Z., Klir, G.J.: Generalized Measure Theory. Springer, New York (2009)
184. Weierstrass, K.: Über die analytische Darstellbarkeit sogenannter willkürlicher Functionen einer reellen Veränderlichen. Sitzungsberichte der Akademie zu Berlin, pp. 633–639 and 789–805 (1885)
185. Wermer, J.: Potential Theory, 2nd edn. Lecture Notes in Mathematics, vol. 408. Springer–Verlag, Berlin (1981)
186. Wiesel, J.: On a Lusin theorem for capacities. Proc. A. M. S. **150**(3), 1351–1360 (2022)
187. Willem, M.: Functional Analysis: Fundamentals and Applications. Springer Science & Business Media, Berlin (2013)
188. Wulbert, D.E.: Convergence of operators and Korovkin's theorem. J. Approx. Theory **1**, 381–390 (1968)
189. Yang, D., Yuan, W: A note on dyadic Hausdorff capacities. Bull. Sci. Math. **132**(6), 500–509 (2008)
190. Yosida, K.: Functional Analysis. Classics in Mathematics, vol. 123. Springer-Verlag, Berlin (1980)
191. Yosida, K., Kakutani, S.: Birkhoff's ergodic theorem and the maximal ergodic theorem. Proc. Imp. Acad. Tokyo **15**, 165–168 (1939)
192. Zhang, D., Mesiar, R., Pap, E.: Pseudo-integral and generalized Choquet integral. Fuzzy Sets Syst. **446**, 193-221 (2022)
193. Zhang, D., Mesiar, R., Pap, E.: Jensen's inequality for Choquet integral revisited and a note on Jensen's inequality for generalized Choquet integral. Fuzzy Sets Syst. **430**, 79–87 (2022)
194. Zhang, D., Mesiar, R., Pap, E.: Choquet type integrals for single-valued functions with respect to set-functions and set-multifunctions. Inf. Sci. **630**, 252–270 (2023)
195. Zhou, L.: Integral representation of continuous comonotonically additive functionals. Trans. Amer. Math. Soc. **350**, 1811–1822 (1998)
196. Zorich, V.A.: Mathematical Analysis II, Universitext. Springer-Verlag, Berlin (2004)
197. Zowe, J.: Sandwich theorems for convex operators with values in an ordered vector space. J. Math. Anal. Appl. **66**, 282–296 (1978)

Index

S. G. Gal, C. P. Niculescu, *Choquet Integral and Monotone Sublinear Operators*, Frontiers in Mathematics, https://doi.org/10.1007/978-3-031-71701-7

FSC
www.fsc.org
MIX
Papier | Fördert gute Waldnutzung
FSC® C083411

Zeitfracht Medien GmbH
Ferdinand-Jühlke-Straße 7
99095 Erfurt, Deutschland
produktsicherheit@kolibri360.de